AN INTEGRATED APPROACH TO LOGISTICS MANAGEMENT

PRENTICE-HALL INTERNATIONAL SERIES
IN INDUSTRIAL AND SYSTEMS ENGINEERING

W. J. Fabrycky and J. H. Mize, Editors

ALEXANDER *The Practice and Management of Industrial Ergonomics*
AMOS AND SARCHET *Management for Engineers*
BANKS AND CARSON *Discrete-Event System Simulation*
BANKS AND FABRYCKY *Procurement and Inventory Systems Analysis*
BEIGHTLER, PHILLIPS, AND WILDE *Foundations of Optimization, 2/E*
BLANCHARD *Logistics Engineering and Management, 2/E*
BLANCHARD AND FABRYCKY *Systems Engineering and Analysis*
BROWN *Systems Analysis and Design for Safety*
BUSSEY *The Economic Analysis of Industrial Projects*
CHANG AND WYSK *An Introduction to Automated Process Planning Systems*
ELSAYED AND BOUCHER *Analysis and Control of Production Systems*
FABRYCKY, GHARE, AND TORGERSEN *Applied Operations Research and Management Science*
FRANCES AND WHITE *Facility Layout and Location: An Analytical Approach*
GOTTFRIED AND WEISMAN *Introduction to Optimization Theory*
HAMMER *Occupational Safety Management and Engineering, 3/E*
HAMMER *Product Safety Management and Engineering*
HUTCHINSON *An Integrated Approach to Logistics Management*
IGNIZIO *Linear Programming in Single- and Multiple-Objective Systems*
MIZE, WHITE, AND BROOKS *Operations Planning and Control*
MUNDEL *Improving Productivity and Effectiveness*
MUNDEL *Motion and Time Study: Improving Productivity, 6/E*
OSTWALD *Cost Estimating, 2/E*
PHILLIPS AND GARCIA-DIAZ *Fundamentals of Network Analysis*
PLOSSL *Engineering for the Control of Manufacturing*
SANDQUIST *Introduction to System Science*
SMALLEY *Hospital Management Engineering*
THUESEN AND FABRYCKY *Engineering Economy, 6/E*
TURNER, MIZE, AND CASE *Introduction to Industrial and Systems Engineering, 2/E*
WHITEHOUSE *Systems Analysis and Design Using Network Techniques*

AN INTEGRATED APPROACH TO LOGISTICS MANAGEMENT

NORMAN E. HUTCHINSON

Florida Institute of Technology

HD
38.5
.H88
1987
WEST

PRENTICE-HALL, INC., *Englewood Cliffs, New Jersey 07632*

Library of Congress Cataloging-in-Publication Data

HUTCHINSON, NORMAN E. (date)
An integrated approach to logistics management.
Prentice-Hall international series in industrial and systems engineering)

Includes index.
1. Business logistics—Management. I. Title.
HD38.5.H88 1987   658.7   86-8116
ISBN 0-13-468976-3

Editorial/production supervision and
  interior design: *Gloria Jordan*
Manufacturing buyer: *Rhett Conklin*
Series Logo Design: *Judy Winthrop*

© 1987 by Prentice-Hall, Inc.
A division of Simon & Schuster
Englewood Cliffs, New Jersey 07632

All rights reserved. No part of this book may be
reproduced, in any form or by any means,
without permission in writing from the publisher.

Printed in the United States of America

10 9 8 7 6 5 4 3 2 1

ISBN 0-13-468976-3 025

PRENTICE-HALL INTERNATIONAL (UK) LIMITED, *London*
PRENTICE-HALL OF AUSTRALIA PTY. LIMITED, *Sydney*
PRENTICE-HALL CANADA INC., *Toronto*
PRENTICE-HALL HISPANOAMERICANA, S.A., *Mexico*
PRENTICE-HALL OF INDIA PRIVATE LIMITED, *New Delhi*
PRENTICE-HALL OF JAPAN, INC., *Tokyo*
PRENTICE-HALL OF SOUTHEAST ASIA PTE. LTD., *Singapore*
EDITORA PRENTICE-HALL DO BRASIL, LTDA. *Rio de Janeiro*

Contents

Preface xiii

1 Logistics and Management 1

Three Categories of Logistics 2
Logistics Management 3
The Drive Toward Integration 4
  Total Cost Analysis, 4
  The Systems Approach, 4
  Customer Service, 6
  Distribution Channels, 6
A Continued Emphasis 8
Integrated Logistics 8
  Operations Logistics, 8
  Systems Logistics, 10
The Mission of Logistics 11
Summary 12
Questions for Review 12

2 The Elements of Logistics 14

Logistics as a System 16
  Logistics as a Support Function, 16
  A Systems Environment, 18

The Logistic Elements 20
   Transportation, 21
   Storage, 22
   Spares and Repair Parts, 23
   Personnel and Training, 24
   Publications, 24
   Test and Support Equipment, 25
   Facilities, 25
Summary 26
Questions for Review 26

3 An Integration of Logistics   28

Logistics Within the Firm 29
   Logistics: The Horizontal Function, 30
   A Potential for Conflict, 32
The Logistics Function 32
   Environmental Considerations, 33
   Logistic Channels, 34
Logistic Operations 35
   The Performance Cycle, 36
   The Marketing Concept, 37
   The Product Life Cycle, 38
Coordinating the Logistics Function 40
   Forecasting, 41
   Material Management, 42
Summary 44
Questions for Review 45

4 Logistic Prerequisites   46

A Balanced Integration 47
Reliability 48
   Elements of Reliability, 48
   Tools of Reliability, 50
   A Failure Rate Curve, 51
Maintainability 52
   Units of Measure, 53
   Net Maintenance Time, 53
Repair-Level Analysis 54
   Repair versus Discard, 54
   Levels of Maintenance, 55
   Repair-Level Decisions, 56
Life Cycle Cost 57
Summary 59
Questions for Review 60

Contents / vii

5 Logistic Connectivity   61

Transportation Modes   62
  Road, 64
  Rail, 66
  Water, 66
  Air, 66
  Pipeline, 67
Legal Groupings of Carriers   67
  Common Carriers, 68
  Contract Carriers, 68
  Exempt Carriers, 68
  Private Carriers, 68
Multimode Systems   69
Regulating the Common Carrier   69
Transportation Rate Structures   70
  Class Rates, 71
  Commodity Rates, 71
  Special Rates, 72
Summary   72
Questions for Review   73

6 Logistic Facilities   74

The Maintenance Facility   75
  Sizing the Maintenance Facility, 76
  Expected Repair Actions, 76
  Workload Determinants, 78
  Personnel Requirements, 80
The Warehouse Facility   81
  The Addition of Place Utility, 83
  Warehouse Functions, 84
  Location Planning, 85
  Economic Justification, 86
  Personnel Requirements, 87
  Material Handling, 88
Summary   89
Questions for Review   90

7 The Logistics of Risk   91

What Is Inventory?   92
  Forms of Inventory, 92
  A Conflict of Functions, 93
  The Functions of Inventory, 94

Inventory Risk   95
    Manufacturing Risk, 95
    Wholesale Risk, 96
    Retail Risk, 96
The Inventory Cycle   96
    Average Inventory, 97
    Pipeline Inventory, 97
    Inventory Performance Cycles, 98
    Adding Pipeline Inventory, 101
Elements of Inventory Cost   103
    Transportation Costs, 103
    Storage Costs, 103
    Obsolescence Costs, 104
    Opportunity Costs, 104
    Order Costs, 104
Production and Inventory Control   105
    Formal and Informal Systems, 105
    Functions of Production and Inventory Control, 105
    Economic Order Quantities, 107
    Order Point, 109
    Discrete Lot Sizing, 110
Summary   112
Questions for Review   113

8   Material Requirements Planning   114

MRP: The Beginning   115
    Time-Phased MRP, 116
    Master Production Schedule, 118
    Capacity Requirements Planning, 119
    Elements of Success, 120
Beyond Material Requirements Planning   120
    Just-in-Time, 121
    A Contrast of Philosophies, 122
    Implementing JIT, 123
    Quality Function, 124
    What Is Kanban?, 125
Summary   126
Questions for Review   126

9   The Era of Customer Service   128

Logistics as a Service   128
What and Who Is the Customer?   130

The External Customer, 130
The Internal Customer, 132
A Service Policy   132
   Capability, 133
   Availability, 133
   Quality, 134
Planning a Customer Service Policy   135
   Least-Total-Cost Design, 136
   Sensitivity Analysis, 137
Implementing a Customer Service Policy   138
Alternatives to Changing the Logistic Structure   139
   Postponement, 141
   Consolidated Shipments, 141
   Reverse Logistics, 142
Summary   142
Questions for Review   143

10   Personnel and Training   144

Training as an Element of Logistics   144
   The Logistics Instructor, 145
   Logistics Instruction, 146
   The Logistics Training Course, 146
Prerequisites to Training   147
   Types of Training, 147
   Student Population, 148
A Training Philosophy   148
   Training Objective, 149
   Writing the Objective, 150
Training Requirements Analysis   150
   Course Length Determinants, 152
   Time Phasing, 153
Preparing the Training Program   154
   Scope of Training, 154
   Course Prerequisites, 155
   Preparing the Instructor, 155
Supporting Course Materials   156
   Training Plan, 156
   Course Outline, 157
   Lesson Guides, 157
   Hands-on Training Guides, 157
   Audio-Visual Aids, 158
   Student Text Material, 158
   Examination, 158
Summary   158
Questions for Review   159

11 Technical Publications 160

The Technical Writer 161
The Technical Publication 161
  Clarity of Presentation, 162
  Organization, 162
  Accuracy, 162
  Adequacy, Relevance, and Effectiveness, 163
The Intended Audience 163
The Scope of the Publication 164
A Necessary Prerequisite 164
  Job Analysis, 165
  Maintenance Items, 166
  Tools and Test and Support Equipment, 167
  Personnel Requirements, 167
  Specific Activity Data, 168
An Overview of the Development Process 168
Design of the Technical Publication 170
An Approach to Publication Design 171
  Front Matter, 171
  Chapters, 171
  Appendices, 172
  Index, 172
  References, 172
Evaluating the Product 173
Summary 173
Questions for Review 174

12 Spares and Repair Parts 175

Sparing Decision 176
Short History of Provisioning 177
Provisioning: The Government and the Firm 178
  Provisioning Process, 179
  Early Provisioning Process, 179
  Logistic Support Analysis and Provisioning, 181
A Provisioning Example 181
Sparing Determinants 183
  SMR Coding, 183
  A Sample SMR Code, 184
  Spares Quantity Determinants, 187
  Budget Constraints, 188
Postproduction Support 188
Postproduction Spares 189
Postproduction Planning 189
Summary 190
Questions for Review 191

13 Test and Support Equipment   192

Test and Support Equipment Categories   193
Levels of Maintenance   193
   Organizational Maintenance, 193
   Intermediate Maintenance, 194
   Depot Maintenance, 194
Requirements for Test and Support Equipment   195
A Test and Support Equipment Program   196
   Selection Process, 197
   Quantity Determination, 199
Logistic Requirements   199
Summary   199
Questions for Review   200

14 Logistic Support Analysis   201

Objectives of LSA   202
Process of LSA   203
   Support Analysis, 203
   Supportability Assessment and Verification Analysis, 204
Application of LSA   205
Logistic Support Analysis Record   205
LSA and the Product   206
Measurement Techniques of LSA   206
Logistic Support Models   208
Life Cycle Cost Analysis   209
   Maintenance Engineering Analysis, 210
LSA and the Deployed System   212
Summary   213
Questions for Review   214

15 Logistics Management and Control   216

What Is Management?   216
   First-Line Management, 217
   Middle Management, 217
   Top Management, 217
Management Thought   218
   Traditional Viewpoint, 218
   Systems Viewpoint, 219
   Behavioral Science Viewpoint, 219
   Contingency Viewpoint, 219
Managerial Power and Influence   220
   Sources of Power, 220
Motivation and the ILS Manager   221

Process of Motivation 221
  Expectancy Theory, 222
  Behaviorism, 223
Logistics Management Philosophy 223
  Management by Objective, 224
  Process of MBO, 224
Control of the Logistic System 225
Earned Value 226
  Progress Measurement, 227
  Purchased Item Measurement, 229
  Milestone Selection, 230
Evaluation of the Report 231
  Performance Indices, 232
  Schedule Conversion, 234
Summary 234
Questions for Review 235

16 The Future of Logistics 237

Inducements to Change 238
The Solution 239

Index 241

Preface

Logistics is rapidly becoming the profession of the future. It has long existed as a fragmented and elusive activity with little recognition beyond the military services and those firms that were a part of the military-industrial complex. Yet, even in this select society, logistics remained an anomaly. The majority of practitioners employed within this field could not agree on a description of logistics, nor could they identify the objectives of their chosen career. The typical logistician toiled daily in his or her portion of the logistics field, caring little about the entirety of this diverse discipline.

This began to change in the latter half of the twentieth century as business realized the potential benefits of logistics. Logistics, properly applied, has the capability of greatly enhancing the profit potential of the firm. The first step in the realization of these benefits is, however, the amalgamation of logistics into an integrated whole. This integration, and control of the logistics function by the integrated logistics manager, is the subject of this book. Support engendered by the logistics function is examined from the viewpoint of the firm as a system. This treatment extends logistics far beyond traditional concepts of material management, internal inventory transfer, and physical distribution. These subjects are properly considered as they are certainly a part of the logistics function. They are, however, supplemented by the remaining elements of the logistics family, such as reliability and maintainability, spares and repair parts, personnel and training, test and support equipment, and technical publications. These are the elements that are required to provide support to the product both before and after the point of ownership transfer. The many elements of logistics are presented in successive chapters throughout the book.

Logistics is unique in that it embraces an astonishing variety of professional disciplines. This compounds the challenge to the integrated logistics manager who must create a package of logistic resources capable of providing optimum support to the product. The problems associated with logistics management, and the associated solutions thereto, are discussed with each of the logistics elements and presented in detail in Chapter 15.

This book is not for the logistics practitioner who is content with working daily in one of the many logistic elements. Each element of logistics represents a sophisticated and dedicated career choice, and it would be a gross absurdity to claim that a single document could present a thorough discussion of each one. Rather, this book is designed for the integrated logistics manager and others who wish to increase their knowledge and understanding of logistics.

NORMAN E. HUTCHINSON
Florida Institute of Technology

AN INTEGRATED APPROACH TO LOGISTICS MANAGEMENT

1

LOGISTICS AND MANAGEMENT

The subject of this book—logistics and the science of management—is unique in that it unites one of the oldest yet most recent enterprise activities with current concepts in management theory and thought. Logistics, as an activity incorporating physical facilities or structures, locations, transportation, inventory, handling, and storage, has been a necessary function since the first commercial endeavor. It is impossible to visualize such an activity that does not demand the inclusion of logistics. The smallest enterprise requires a facility at some location, an inventory with which to produce a product, and transportation for the acquisition of production material and the subsequent physical distribution of finished goods. These activities necessitate handling and storage as a result of the acquisition, production, and distribution phases of the enterprise.

Where, then, is the newness of logistics? Logistics is a support element of the enterprise, and its newness stems from an integrated approach that began to surface during the 1950s. A working definition of logistics during this period of emergence would have considered logistics as managing the movement and storage of

A. Material into the enterprise
B. Goods in process in the enterprise
C. Finished goods from the enterprise

This concept of logistics has been subject to varying degrees of acceptance and has produced varied results. What are the underlying reasons behind these performance variations? Let us attempt to evaluate the cause or causes by analyzing each action as identified in the definition given above.

A. "Managing." *Managing* is controlling the logistic activity in relation to an overall plan or strategy.
B. "Movement." Raw materials must be moved from the supplier (or suppliers) into the facility. These same raw materials must be moved through the facility as they are converted into finished goods. The finished goods must be moved to the customer.
C. "Storage." Raw materials must be stored until needed for production; work in process must be stored during various stages of the production process; finished goods must be stored until requested by the customer.

The definition above accurately defines the logistic process but is too restrictive under the modern concept of an integrated approach to logistic support. Logistics, as a support element of the enterprise, incorporates activities beyond those identified in the preceding definition. Consider, for example, the enterprise that introduces a new product or a significant change in an existing product. Support of the new product or feature may require training, the development of technical publications, and the acquisition of spare parts and special tools or repair (test) equipment. These items also belong under the umbrella of logistics; thus our definition must be revised to include them. This, however, would result in a rather cumbersome expression. A simplified definition would be: *"The process of having the right quantity of the right item in the right place at the right time."* The management of these integrated logistic support (ILS) activities is thus the task of assuring that these objectives are achieved within acceptable resource limits.

Three Categories of Logistics

Integrated logistics support, when properly understood and applied, can provide the means to identify and resolve many logistics problems, frequently before they develop. But what do the terms *logistics* and *integrated* mean in this context?

Logistics, in the broadest sense of the word, can be considered as a scope of activity comprised of three major areas or subsets: (1) subsistence logistics, (2) operations logistics, and (3) systems logistics.

Subsistence logistics is concerned with the basic necessities of life such as food, clothing, and shelter. At any given time, within any given environment, subsistence logistics is relatively stable and predictable. Men and women, as rational beings, know within very narrow limits what is needed, how much is needed, where it is needed, and when it is needed. Subsistence logistics is the primary activity of primitive societies and is an essential ingredient of an industrial society. It provides the foundation for operations logistics.

Operations logistics extends beyond the bare necessities by incorporating systems that produce the luxuries or niceties of life. By definition, operations logistics incorporates the raw material required by the enterprise in the production process, the means of utilizing that material during the production process, and the distribution of finished goods resulting from production. This category of logistics is also relatively constant and predictable. All enterprises, from the automobile manufacturer to the fast-food chain store, can determine the quantity of materials and the resources needed for its production with a high degree of accuracy. Operations logistics cannot, however, determine when a component of the enterprise is going to break down, what will be required to repair it, or the duration of the repair activity. Operations logistics, which is concerned with the movement and storage of materials into, through, and out of the enterprise, provides the foundation for systems logistics.

Systems logistics incorporates the resources required in keeping a system in operating condition. These resources, or logistic elements, are spares and repair parts, personnel and training, technical publications, test and support equipment, and facilities. A well-designed integration of these logistic elements is critical when, for example, repair instructions describe one method of repair and tools are developed for another method. Thus repair may be impossible. The critical nature of this integration is recognized in the Department of Defense (DOD Directive 4100.35) definition of Integrated Logistic Support, which states,

> Integrated Logistics Support is a composite of all support considerations necessary to assure the effective and economical support of a system throughout its existing life. It is an integral part of all aspects of the system and its operation. Integrated Logistics Support is characterized by harmony and coherence among all logistic elements.

Harmony and coherence among logistic elements are the primary objectives of logistics management.

Logistics Management

Logistics management is concerned with the development and implementation of a methodology for ensuring the efficient and cost-effective attainment of logistic objectives. Logistic objectives, under the simplified definition previously presented, are defined as having the right quantity of the right item in the right place at the right time. The "right item" may be identified as raw materials, finished goods, trained personnel, spare parts, or any of the other logistic elements.

The application of logistics management principles frequently represents a compromise among conflicting interests within the enterprise. Consider, for example, the acquisition of raw materials for movement into the facility. The buyer of those materials may rightfully assume that a cost-effective approach is to purchase vast quantities, thus obtaining the lowest per-unit price. This parochial outlook may, however, be in direct conflict with the manager of inventory, who may very well incur large costs in the storage of such huge quantities. Integrated logistics management is responsible for resolving such conflicts in a manner that is both cost effective and efficient from the viewpoint of the enterprise as a whole.

Logistics, within the concept of integrated logistic support (ILS), requires an ILS manager. The ILS manager is the "interface" among logistic elements, whose objective is to design a package of logistic resources that is characterized by harmony and coherence.

The Drive Toward Integration

The drive toward an integrated approach to logistics can be traced to the 1950s when a volatile economic climate led to a squeeze on profits, thereby leading to a massive search for cost-reduction measures. Prior to this time, logistics was implemented on a fragmentary basis, and there was little realization that integrated logistics could lead to improved performance and cost reductions. An early realization of the benefits of integrated logistics was hampered because many of the quantitative techniques in use today were impractical prior to the advent of computers as a management tool.

The decade between 1956 and 1965 witnessed a quantum jump in the development of an integrated logistic support concept. This progress can be directly related to major developments in the areas of

A. Total cost analysis
B. Systems approach
C. Customer service
D. Distribution channels

Total Cost Analysis

Total cost analysis was first applied to logistic economics during the mid-1950s in an attempt to evaluate the cost of logistic problems. Total costs refer to all costs associated with a logistical mission—from the initial concept exploration phase to the end of that product's useful life—and is called the *life cycle cost* (LCC). Thus this total cost concept reduces the entire product life span to a common denominator—dollars and cents.

The Systems Approach

The manager who views an enterprise (a business concern or firm) as an integration of functional areas working together toward a common goal is considering that enterprise as a system. Various areas within the firm are integrated by the flow of resources, with each area depending on the others for survival. The firm then becomes a physical system just as a machine such as an automobile is a system.

A systems viewpoint or orientation requires acknowledgment of the environment in which the firm operates. The firm not only draws resources (raw materials) from the environment, it also makes a contribution to that environment through the distribution of finished goods. A systems orientation is critical to the systems approach.

The word *system,* as used in the context of logistics, identifies a grouping of logistic elements or activities that are integrated through the common purpose of attaining an objective. Regarding logistic activities as a system is an abstract method of thinking, but this abstraction prevents the logistics manager from getting lost in the details by emphasizing the importance of harmony and coherence among logistic elements.

The firm as a system is depicted in Figure 1.1. In this view, the input (raw materials) undergoes a transformation process to produce an output (the finished good). The control mechanism (logistics management for ILS activities) monitors the system and regulates its operation to properly execute the transformation process.

A manufacturing concern such as an automobile plant is an illustration of the systems concept. The plant requires an input of materials with which to produce the product or finished good, an automobile. The material input to the plant is referred to as raw materials, although a moment's reflection reveals that a raw material input to one firm is the finished good output of another firm. Iron ore, for example, is the raw material input to a steel mill; yet it is the finished good output of a mining operation.

The automobile plant utilizes the input material during the transformation process, wherein the multitude of component parts (raw materials) are combined to produce an automobile. The control element (manager) assures that the assembly line moves at the correct rate and that the many parts arrive at the assembly line in the right quantity and at the right time. Thus, the plant functions as a system wherein incoming raw materials undergo a transformation process resulting in the finished product.

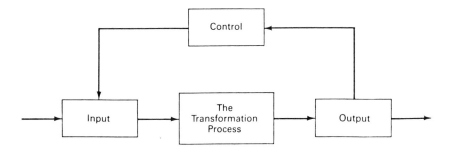

Fig. 1.1. The Firm as a System

Another example of the systems process is a class of students in a school or university. The raw material input (the student) undergoes a transformation (the learning process) resulting in a finished good (a student who passes). Here again, it is appropriate to recall that the incoming student represents a raw material as well as a finished good (having completed a previous class).

The systems approach to integrated logistics dictates that the manager view the firm as a whole, not as individual units. This frequently creates a need for compromise among traditional management functions within the firm. For example, the marketing manager may desire a wide diversity of products to permit timely reaction to potential customer demands. The production manager, on the other hand, prefers long runs of the same product to minimize lost time and inefficiency due to changes in the production process.

Customer Service

The 1960s witnessed an expansion of the integrated logistics concept with a concurrent shift in managerial emphasis from cost to customer performance. This shift resulted in a more realistic approach to logistic activities and their interrelationship with manufacturing and marketing. Various logistical systems, each offering different levels of operational support and resource commitments, could now be developed to support a variety of manufacturing or marketing plans. Logistics management was given the task of developing and implementing a logistical system capable of attaining the required degree of customer service at the minimum possible cost.

Distribution Channels

The growing awareness of logistics as an integrated activity was coupled with an increased realization of the importance of distribution channel arrangements to the logistic system. Significant and frequently unnecessary logistics costs or an impairment of customer service may be incurred as a result of individual enterprises engaging in a *channel arrangement*.

Early analyses of the distribution channel were centered on location. The enterprise should be located either near a source of raw materials or at a location that assured uninterrupted and economical transportation of raw materials from their source to the enterprise. Conversely, the ready availability of distribution channels for the movement of raw materials into the enterprise permitted an alternate approach whereby the firm could be located near its customers. This emphasis on location virtually ignored the impact of time on the logistic system.

The analysis of distribution channels and the logistic system received added impetus with the recognition that logistic activities and responsibilities frequently continue beyond the point of ownership transfer. The federal government, to a significant degree, contributed to this recognition through the forced recall of automobiles and interpretations of implied product warranties.

Possibly the greatest achievement of this research into the distribution channel was the integration of time and location. This approach offered a more balanced view of logistics by taking the integration of spatial and temporal forces into consideration.

A Continued Emphasis

During the latter part of the 1960s, the logistics manager came to be recognized as a viable entity within the business community. This "new" manager was now equipped with a variety of segmented, yet theoretically sound, approaches to logistics planning. The previous period of exploration was replaced with a period of testing: Does logistics theory really work?

The initial period of testing was characterized by an emphasis on one of the two primary aspects of operations logistics. Therefore, within a given enterprise, logistic activities were concentrated in either the marketing arena or the materials management arena.

The emphasis of logistics placed upon the marketing arena evolved into physical distribution management as an integrated approach to the movement of finished goods from inventory to the customer. Logistical support of customer orders was the main thrust of this approach.

In contrast to the marketing approach, the application of logistics to materials and the subsequent manufacturing process led to the emergence of materials management and the material manager. Material management emphasized the orderly flow of materials, both into and through the enterprise, with time-phased delivery being scheduled in accordance with need.

The period of testing was followed by an interval of changing priorities engendered by economic events of the 1970s. These changes were highlighted by the energy crisis, which forced logistics to face the need for drastic improvements in productivity. This was compounded by the high visibility of transportation and storage as large energy users.

Another characteristic of the 1970s contributing to this shift in priorities was the recession during the early part of the decade. The recession led to a period of "stagflation" (*stag*nation in the growth of the gross national product coupled with simultaneous in*flation*), which emphasized the need for additional economies within the enterprise.

As a direct result of the changing economic climate of the 1970s, enterprise priorities dramatically moved away from the mere production of products to the service of demand. The firm could no longer simply produce products to satisfy an apparently insatiable craving for more by the consumer. This shift resulted in the immediate growth of materials management as a profession.

The federal government and the vast military-industrial complex had significant impact on the development of an ever-increasing logistic capability. U.S. government actions primarily influenced systems logistics, but there was also considerable impact on operations logistics.

Government influence in logistics became a reality with the realization that the costs associated with logistical support throughout the useful life of a system (from concept exploration to retirement) frequently exceed initial acquisition cost. Governmental recognition of the importance of logistics can be seen by the weight accorded a logistics capability in their Requests for Proposals (RFPs) offered to private industry. (RFPs are the vehicles by which the Federal Government asks the private sector to submit proposals for providing goods and services to the public sector.) It is not unusual for logistics to equal or exceed the weighting accorded all other factors when determining who should be awarded a contract.

Integrated Logistics

The trend toward integration in logistics surfaced with the belief that integrated performance produces results that are superior to those produced by individual functions acting in relative isolation. The current challenge is to recognize and act upon the very real interdependence among logistic elements. A systems orientation to logistics forces this recognition and requires the logistics manager to develop logistic strategies that are properly designed to support the objectives of the enterprise. This approach must acknowledge the variable and contrasting goals of the various functions that make up the system (enterprise). Finance, as a vital component of the system, may have the objective of minimizing the amount of cash tied up in assets such as a large inventory. The manager of inventory, however, may prefer a large inventory, thus permitting a rapid response to requests for service.

The role of the integrated logistics manager is to resolve these types of conflict. Management intervention must be directed toward a compromise, wherein the objectives of functional units within the enterprise may have to be reduced (suboptimized) for the benefit of the firm.

Integrated logistics combines operations logistics and systems logistics into an interrelated whole that strikes a balance between reasonable performance levels and realistic cost expectations.

Operations Logistics

Operations logistics is an amalgamation of two separate but related activities— logistic operations and logistic coordination. *Logistic operations* is concerned with the movement and storage needs of the enterprise, whereas *logistic coordination* is related to the identification of those needs.

Movement and storage needs may be related to the systems concept of the firm as depicted in Figure 1.1. The input component of the firm bears responsibility for the acquisition of resources or raw materials that are necessary for the production process. The acquisition of these materials must be preceded by the selection of a source or sources. Factors entering into this selection process include

A. The capability of the source in meeting the demands of the firm
B. The location of the source with respect to the firm
C. The availability of suitable transportation channels
D. The reliability of the source

Source selection must be followed by the movement of material into the firm. Material management must incorporate provisions for the storage of this material from the time it is received until it is used in the production process.

Material received as an input to the firm may now be used for the production of finished goods. As input material is transformed into the finished product, semifinished goods exist within the firm. These semifinished goods are controlled by a process called *internal inventory transfer*. This function also entails a storage requirement as partially finished goods may have to be stored until needed during subsequent stages of production.

The culmination of the production process is the finished good. These products, excluding the rare firm that builds to order, require storage until requested by the customer. Upon sale, physical distribution of the product becomes still another logistic concern. Logistic responsibility does not, however, end with distribution to the customer. Consider, for example, the product sold under warranty that must be returned for service or the previously referenced mandated recalls of defective automobiles. This activity represents a reverse flow through the distribution channel.

The companion of logistic operations is *logistic coordination*. While logistic operations is concerned with the movement and storage of items into, through, and out of the enterprise, logistic coordination is concerned with the identification of the need for those logistic activities. As in logistic operations, the identification of movement and storage needs may readily be reviewed through reference to the systems model of the firm (see Figure 1.1).

The output component of the systems model relies upon *product market forecasting* to provide advance estimates of the demand for the goods of production. Forecasting, or the process of predicting future events through a systematic analysis of relevant data, must extend beyond a simple projection of the quantity required over a given period of time (usually one year). Projections of future demand must also consider potential variations in that demand. As an example, there is a relatively small but steady year-round market for toys to accommodate birthdays, special occasions, and general gift giving. There is, in addition, a known and predictable increase in this demand prior to the Christmas holidays.

The input component also relies upon product market forecasting to provide an indication of the need for materials that enable the transformation function to respond to the demand for finished goods. The movement and storage of materials into the firm lag behind the finished good requirement by an amount of time approximately equal to the transformation process. Material must, however, be requested by the firm in advance of the need date. The magnitude of this advance time is approximately equal to transportation (pipeline) time plus order processing and handling time. Proper timing of this activity is frequently referred to as *material requirements planning* (MRP). (MRP may also be expanded to encompass the entire production process, in which case it is referred to as *manufacturing resource planning.*)

Overriding the input-transformation-output process of the firm is an activity referred to as *operational planning.* Operational planning attempts to integrate the capabilities of the firm with the desires of management. The production capabilities of any firm are finite and, in the short term, can only be expanded within narrow limits (through the judicious use of overtime or by going to additional shifts, for example). Management must recognize these limits and refrain from overcommitting the resources of the firm.

Systems Logistics

Systems logistics incorporates the elements of operations logistics (the movement and storage of materials, work in process, and the finished good) and extends the process to encompass those logistic elements providing support subsequent to product distribution. These logistic elements include spares and repair parts, personnel and training, technical publications, test and support equipment, and facilities.

As an illustration of logistics support following production, consider the ever-increasing complexity of the consumer marketplace. Take, for example, the video recorder or the personal computer industry. Is it reasonable to assume that the responsibilities of the enterprise terminate with physical distribution of the finished good to the consumer? Most assuredly not! The firm must, either by itself or through third-party intermediaries, provide for the support elements of systems logistics.

A major and significant difference between operations logistics and systems logistics is the predictability of operations logistics as opposed to the random nature of systems logistics. The demand for any given product may be predicted with reasonable accuracy through the application of sophisticated market forecasting techniques. Demand for the product is directly related to the demand for materials used in the creation of that product. This same predictability is not true for systems logistics, however. The failure of a finished good is random in nature, and predictions within this arena must rely on the mathematics of probability and statistics.

The basic support element of systems logistics is the spare part. Without parts to replace those that have failed or are suspected of failure, an item quickly becomes inoperable. Failures create the demand for repair parts, yet failures are random events. Thus the derived demand for spares is also random. The random nature of the spares process does not, however, obviate the need for procurement and distribution of discrete quantities. How, then, are acquisition quantities and distribution schedules to be determined?

The quantity of spares required to support a given item is determined by (1) statistical analysis of the expected rate of failure, (2) the time required to repair the failed item, (3) the total number of items being supported, and (4) the average time between the request for a repair part and the receipt of that item (transportation or pipeline time plus processing and handling times). The analysis must be performed in advance of need to permit ordering and distribution in time to meet the need created by the failure.

Another element of systems logistics is the technical publication. Publications that efficiently and effectively support the finished good must be developed. Depending upon the complexity of the product being supported, publications may range from a one-page brochure to a multivolume technical reference library. Publication content may range from simple operating instructions to detailed maintenance procedural data incorporating instructions for alignment and adjustment, malfunction identification and isolation, and restoration of service. Maintenance instructions, when included with the technical publication, must incorporate parts identification and the proper use of tools and test equipment. All instructions and procedural data must be consistent with the capabilities of the expected audience.

The principal support activity involves adjustment, service, and repair of the item being supported. This activity is facilitated through the use of test and support equipment that has been selected or designed to accommodate the derived requirements. Test and support equipment may be, and frequently are, more complex than the supported item. The combination of selection, quantity, and timely delivery of this test and support equipment is yet another element of systems logistics.

Personnel may well represent the most expensive element of systems logistics. This cost may, however, be minimized through the careful selection and training of support personnel. Training must be developed that is consistent with the product, the technical documentation, the repair parts inventory, and the test and support equipment.

The last element comprising systems logistics is the facility. The package of logistic resources cannot be considered complete until the facilities required in providing logistic support have been acquired. Facilities may be new or a modified structure. First, a survey of existing facilities is needed. The results of the survey are then compared with logistic support requirements. Any shortfall between what is available and what is needed must be compensated through cost-effective utilization of resources.

The Mission of Logistics

The *mission of logistics* may be defined as the development of a logistic system that meets the desired logistic performance objectives at the lowest possible price. Within this context, the challenge of logistics is to establish a balance between performance and cost that optimizes the goals of the enterprise.

SUMMARY

The integrated logistic support concept is a relatively recent activity of the enterprise, even though logistics has been a necessary specialty for thousands of years. Recent quantum jumps in the logistics arena have been stimulated by a need for improved performance at a lower cost. This was made possible by a vast array of technological advances. The application of management disciplines holds the promise of enormous payoffs in the decades ahead.

Subsistence logistics encompasses the basic necessities of life. Operations logistics incorporates subsistence logistics and adds the niceties of life. Operations logistics is concerned with the movement and storage of materials into the firm, semifinished goods through the firm, and finished goods from the firm.

Systems logistics encompasses operations logistics and the elements required in providing support following product distribution, such as spares and repair parts, personnel and training, technical publications, test and support equipment, and facilities.

Logistics management is the art of using logistic resources to ensure a cost-effective and efficient attainment of logistics objectives. Logistics objectives may be defined as having the right quantity of the right item in the right place at the right time. Logistics management may be facilitated by the systems approach, wherein the firm is viewed as a system. The firm, when viewed as a system, incorporates an input, a transformation process, and an output. It is made up of interrelated and interactive components and functions as a unified whole.

This first chapter has introduced the concept of logistics and the various logistic elements that work together to provide the resource identified as integrated logistic support. Subsequent chapters provide a detailed treatment of each logistic element and elaborate on the systems concept as the integrating force behind integrated logistic support.

QUESTIONS FOR REVIEW

1. Discuss recent advances that have contributed to the growth of integrated logistic support as a major management activity.
2. Define the life cycle cost (LCC) concept.
3. What is meant by suboptimization of goals with respect to the components of a system?
4. Differentiate between material management and product distribution.
5. Discuss training and why it is defined as one of the elements of systems logistics.
6. What is the difference between operations logistics and systems logistics?
7. What is wrong with the statement that the objective of logistics is to obtain the lowest possible cost?
8. Discuss the concept of material requirements planning.

9. The text defined the objective of logistics as having the right quantity of the right item in the right place at the right time. How does this apply to spare parts? How does it apply to a facility?
10. What is logistical performance?

2

THE ELEMENTS OF LOGISTICS

Chapter 1 introduced the concept of logistics and introduced three discrete subsets: subsistence logistics, operations logistics, and systems logistics. Are there, in fact, three discrete and identifiable subsets of the whole we now call *logistics?* The answer is a qualified no! To expound on this, let us consider what we have identified as three subsets of logistics from two different vantage points.

The first viewpoint is the introductory approach taken in Chapter 1 where logistics is seen as a horizontal array. This approach depicts subsistence logistics as being the province of a more stressful period where the total of life's activities are compressed into a continuous struggle for existence. (Subsistence logistics, you will recall, provided only for the necessities of life.) The next subset, operations logistics, indicated a less stressful period by providing for the niceties of life in addition to the necessities provided by subsistence logistics. Operations logistics is concerned with the movement and storage of (1) raw materials into the firm, (2) semifinished goods through the firm, and (3) finished goods from the firm. Prior to the 1950s, operations logistics was the only type of logistics of interest to the majority of firms. This interest quickly became more sophisticated as tremendous advances took place in technology and computers became as common as typewriters throughout industry.

These advances, characterized by acceptance of the computer as a management tool, initiated the move toward an integration of logistics. This was followed in short order by increased concern for the product beyond the point of distribution. Industry, faced with government intervention and an increasing array of high technology products, could no longer abdicate responsibility with the transfer of ownership. These events led to the development of an area of logistics concerned primarily with support of the finished good following physical distribution—systems logistics.

Figure 2.1 illustrates the concept expressed in the preceding paragraphs, wherein logistics is viewed as a horizontal array. First there is subsistence logistics, where entire lifetimes are consumed in an unending struggle for existence. The individual in this condition possesses neither the means nor the desire to produce a finished good for distribution to others. Virtually all effort is expended toward survival of the individual. What logistics there is consists of the gathering of materials such as food, clothing, and shelter to support the continuation of life.

Subsistence logistics can and does exist as an independent activity. This, however, is temporary. As conditions improve, there is a perceptible move toward specialization. The individual possessing a talent for the building of chairs begins to expend energies toward that end. Production may be in excess of need, with the resulting surplus being available as finished goods to others.

The chair builder now needs additional raw materials for production. Chairs exist as raw materials awaiting transformation, in various states of assembly as semifinished goods, and as finished goods awaiting delivery to a customer. The elements of operations logistics (the movement of materials into, through, and out of the firm) are now in place. Operations logistics cannot, however, exist alone: It must be based on a foundation of subsistence logistics.

In this society there is no need for logistics to extend beyond the point of ownership transfer. The chair (product) that breaks is repaired by the new owner. The age of specialization, however, leads inexorably to the Industrial Revolution, with a concurrent increase in the quantity, variety, and technological content of the finished good. The purchaser can no longer repair the product. Repair may require training, special tools and test equipment, an inventory of repair parts, and perhaps a special building or facility. Support subsequent to ownership transfer becomes a critical issue. We thus enter into the era of systems logistics.

Systems logistics can only exist on a foundation of subsistence logistics and operations logistics. This leads to the second, and more appropriate, viewpoint of logistics as illustrated in Figure 2.2. The three subsets of logistics are now depicted as a pyramid with a given logistic subset being supported by all lower-level subsets. This representation clearly reveals that operations logistics incorporates and expands upon subsistence logistics and that systems logistics, in turn, incorporates and expands upon operations logistics. This approach most nearly approximates the real world and is the one followed throughout the remainder of this text. This is not intended to imply that each firm must incorporate all logistic elements or resources.

Fig. 2.1. Logistics: A Horizontal View

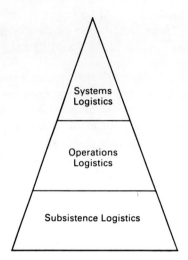

Fig. 2.2. Logistics as a Pyramid of Subsets

A steel mill, for example, requires the movement of raw material (iron ore) into the firm; steel exists throughout the mill in various states of the final product; the finished good (steel) must be stored prior to being distributed to the customer. There is little need for this firm to develop a program of logistics to provide support following product distribution. The steel mill in this example incorporates the elements of operations logistics with an almost total lack of concern for systems logistics.

Logistics as a System

A *system* may be defined as a grouping of items structured so that it fulfills some objective or satisfies a need. A single item cannot be a system: A single board is not a system, but it can be grouped together with other boards to form a fence, which is a system. In like manner, the elements of logistics (or logistic resources) must be integrated together to fulfill the logistic objective of having the right quantity of the right item in the right place at the right time.

Logistics management bears the responsibility of grouping together or integrating logistic elements so that they best meet logistic objectives. The logistics manager is the focal point for all logistic activity and must strive to meet the objective of designing a "package" of logistic resources that is characterized by harmony and coherence.

Logistics as a Support Function

Logistics is a support function in that it exists only to provide support for other components of the firm. Logistics supports the production process (operations logistics) and the product following the transfer of ownership (systems logistics).

This does not imply that the production process does not include elements of systems logistics or that support following ownership transfer does not include elements of operations logistics. In reality, the logistics subsets are intermingled with emphasis frequently being placed on one subset over the other.

Consider, for example, a move toward an increased use of robotics in the manufacturing process. These very elaborate machines may require training to develop the skills needed for their operation and repair. The repair may require technical documentation, special tools and test equipment, and possibly a repair facility—all elements of systems logistics. From another perspective, consider the repair facility dedicated to providing support following transfer of ownership. To effect repair, the facility must maintain an inventory of repair parts, an inventory that must be restocked as parts in storage are consumed during the repair process. The movement of repair parts into the facility and their subsequent storage until required for repair constitute elements of operations logistics. This natural intermingling of logistics renders discussions of operations logistics versus systems logistics superfluous. Thus there is simply logistics, and the elements of logistics are

A. Transportation (movement)
B. Storage (inventory)
C. Spare and repair parts
D. Personnel and training
E. Publications
F. Test and support equipment
G. Facilities

This does not imply that a firm or enterprise must employ all logistic elements. Any given enterprise may employ various combinations of the logistic elements or all elements with varying degrees of emphasis.

Figure 2.3 is another representation of the firm as a system. This diagram extends the system concept by showing the relationship between the activities of the

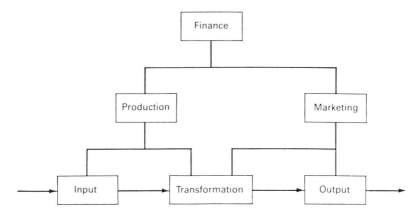

Fig. 2.3. The Firm as a System of Functions

firm and the operating functions of production, marketing, and finance. *Production* is responsible for the transformation process that creates the finished good and the incoming materials that enable its creation. Note that production includes the movement and storage of materials and the semifinished good. *Marketing* is responsible for product sales and incorporates the storage and distribution of finished goods. *Finance* is over the production and marketing functions, as it is the function that provides (and accounts for) the financial resources necessary to sustain continued operation.

A fourth operating function, logistics, is now added to the firm. Logistics, however, supports the other functions as illustrated in Figure 2.4. Production is supported through management of the movement and storage of raw materials into the firm and semifinished goods through the firm. Marketing is supported through management of the movement and storage of the finished good. Logistic support to production and marketing may also entail training, spare and repair parts, or any of the other logistic elements. Production, for example, could require training of employees in the operation and repair of machines used during the manufacturing process. Marketing, on the other hand, could require training for customers in the operation of the purchased product. Logistics may also play a coordination role with the finance function to help resolve any conflicts among functional units of the firm because of a variation in priorities.

A Systems Environment

In Chapter 1 we used the example of a university class as a system: Incoming students are the raw material; class sessions comprise the transformation process; and passing students are the finished good. The class does, in fact, represent a

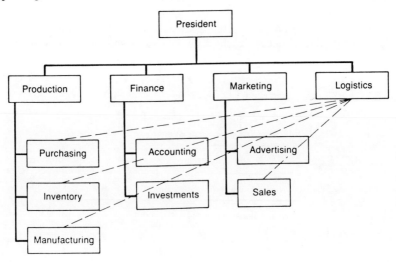

Fig. 2.4. Four Operating Functions of the Firm

system, but it is also a subsystem of the university (a much larger system). The university accepts incoming students as raw material, subjects them to a certain number of years of study (the transformation process), and eventually graduates them (the finished good). The university is, nonetheless, also a subsystem of the education system as a whole. This analogy also holds true for the firm, which is a system as well as a subsystem of the larger system. For example, an automobile plant is both a system and a subsystem of the larger automobile industry.

Changing our viewpoint to look into the firm, it is apparent that operating functions of the firm are likewise both systems and subsystems. The financial system within the firm is also the financial subsystem of the firm. Logistics, as one of the four operating functions of the firm, is also a system within the firm and a subsystem of the firm. Transportation (of raw material into the firm, for example) represents one of the logistic inputs to the logistics system. Integrated logistic support (ILS) management directs the logistic transformation process to obtain the finished good, a package of logistic resources that is characterized by harmony and coherence. Figure 2.5 illustrates logistics as a system.

The systems concept stresses a total integration of all components toward the accomplishment of a predetermined objective. The objective of a given enterprise may be to attain the lowest cost of operation, to provide superior customer service, to maximize profit, or any other objective established by management. In this context, the specific objective of any operating function is subordinate to the goals of the firm. All components must work in concert toward the attainment of the objectives of the enterprise.

This may require that one or more of the operating functions reduce (sub-optimize) an objective for the good of the enterprise. Consider, for example, the manager of an assembly line (part of the production function). An objective of

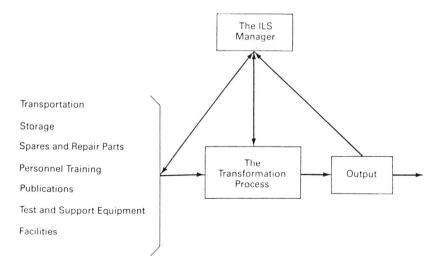

Fig. 2.5. The Logistic System

maximizing efficiency would be best served by long, uninterrupted runs of a standard product with a large inventory of raw materials to avert the possibility of interrupting the manufacturing process. The resulting large output of a single product may be counter to the objectives of a firm desiring to maximize product availability by offering a wide range of diverse products to the consumer. Logistics is vitally interested in the interaction between functions of the firm and the trade-offs that may be necessary in meeting the objectives of the firm.

Consider the following example in which the management objective is wide-ranging product availability. Logistics must consider the logistic elements of transportation and facilities and the derived requirement for maintaining an inventory. Of course, the assembly line manager desires a large inventory of raw materials to assure uninterrupted product runs. Logistics must arrange for the transportation, storage, and management of this inventory. Concerning physical distribution, logistics must also store a large variety of products, maintain an accurate inventory of what is available to assure rapid response to customer requests, and provide transportation for that product.

The Logistic Elements

Logistic elements consist of those activities or resources that represent inputs to the logistics system. Logistic elements, when integrated into a coherent and harmonious whole, support the objectives of the firm. The elements of logistics make up the *logistic function,* which, although essential to cost-effective and efficient operation of the firm, has only in recent times been identified as a separate activity, yet one that becomes meaningless if isolated. Logistics must function as a totality. It must operate in concert with other operating functions of the firm and with external influences that, to some degree, impact the day-to-day operations and the continued existence of the enterprise.

The firm or enterprise is not and cannot be self-sufficient. The corporate design is shaped and molded by the laws and regulations enacted by the political system. The product, the prices charged for the product, and the geographic area served are influenced by the specific industrial system in which the firm operates. The supply of raw materials and the physical distribution of the finished good are controlled by the transportation system. The construction of new facilities and capital improvements are controlled by the financial system. The logistics manager must recognize these relationships and act upon both actual and potential impacts to each of the logistic elements.

This is not to say that industry has relegated all logistic elements to the managerial control of a single integrated logistic support manager. Although strong arguments may be set forth in favor of such an approach and tentative movement in this direction may be observed throughout industry, in actuality the various logistic elements are typically divided among several managers, none of whom necessarily bears a logistics title. For example, the logistic element of transportation may be under the control of a material manager in the production function with assigned responsibilities for incoming materials only. Transportation with regard to physical distribution of the finished good may be the responsibility of another manager within the marketing function. This distribution of responsibilities does not obviate the need that each manager must control the logistic element in concert with the concept of integrated logistic support.

Previously we have listed the logistic elements as transportation, storage, spares and repair parts, personnel and training, publications, test and support equipment, and facilities. These elements do not, however, represent the total logistic task. Subelements within each logistic activity must be considered in developing and understanding and appreciation of the complete logistics arena. The remainder of this chapter presents an overview of each logistic element and introduces associated subelements.

Transportation

The need for the logistic element, transportation, closely parallels the seemingly unending move toward specialization. As a firm narrows its focus to a single product or a group of related products, it becomes increasingly dependent upon the external environment for survival. No enterprise can be self-sufficient. Each member of the industrial system, from the smallest enterprise to the largest corporation, must rely on other firms to provide the materials necessary to sustain operation. Transportation provides the means of moving that material from the source to the enterprise where it is used during the transformation process. The transformation process creates the finished good, and transportation is again required to provide physical distribution of that product. Transportation, the movement of materials into the firm and the physical distribution of finished goods from the firm, represents the lifeblood of the enterprise.

Transportation ranges from the random process whereby the television technician relies upon product breakdown to trigger a material input (the defective television) to highly orchestrated scenarios such as the shipment of crude oil from a well in the Middle East to refineries in the United States. Yet transportation, both into and out of the firm, remains one of the least understood areas of the industrial system.

The importance of transportation as a logistic element may be realized by considering the effects of the oil shortage during the early 1970s. This shortage resulted in huge price increases to the energy user, which in turn created tremendous increases in the cost of doing business. The increase was particularly severe for those firms in the transportation industry. Increased costs led to increased charges for transportation, which ultimately led to higher prices for virtually every product available in the marketplace.

An enterprise rarely, if ever, provides for all its transportation needs through the use of internal resources. Rather, this logistic channel is composed of a variety of independent enterprises acting in concert to deliver material into the firm or finished goods from the firm. The independent enterprise (business) deriving a profit from the movement of material and finished goods is referred to as an *intermediary specialist*. A given logistic channel may be made up of several intermediary specialists, with the gains or losses of each member determined by the success (or lack thereof) of the overall channel.

Transportation may also exert considerable influence upon the location selected for a given enterprise. The ready availability of water or rail transportation, for example, would permit the firm to be located a considerable distance away from the source of raw materials; whereas a lack of these transportation resources would require that the firm be located near the raw material source. Physical distribution of the finished good may also impact the selection of a firm's location. The availability of cost-effective and efficient logistic channels of distribution permits the selection of a firm's location with little consideration of the market area being served.

A relatively recent innovation in the logistics channel is reverse logistics or the physical distribution from the customer to the firm. The frequency of product recalls and the increasing demands of a more sophisticated consumer stress the importance of this reverse channel traffic.

In summary, transportation logistics must include the firm, the suppliers of the firm, the market area served by the firm, the needs of the firm, and the transportation capabilities of available logistic channels to properly assess and integrate transportation with the other logistic elements.

Storage

Storage and the activities associated with it represent a vital logistic element and a critical link in the logistic channel. Storage is required in all phases of the production process, from incoming raw materials to the finished product. Storage requirements, such as the type of facility and the quantity of material to be stored, represent variable factors which must be considered. For example:

A. A supplier to the firm does not produce the needed material on demand. Rather, a supply of material is produced in expectation of a later purchase. Commitments for future sales and potential future sales weighted by the probability of occurrence determine the amount of storage space and the type of structure required for storage.

B. The firm does not purchase materials required during the transformation process on an as-needed basis. (The exception to this is the concept known as "just-in-time" which will be discussed later in this book.) Market projections indicate future demand for the product, thus establishing the magnitude of the transformation process. This, in turn, defines the quantity that is prudent to store at the firm. Uncertainties in the labor market such as the possibility of a strike at the supplier may vary the amount of material a firm wishes to store. This eventuality may very well be handled through temporary arrangements with an intermediary specialist offering storage facilities for hire.

C. A firm rarely produces a product to order. Conventional business practices dictate production in excess of current need as based upon market projections and anticipated growth. Finished goods that are unsold remain the property of the firm and must be stored in a warehouse or storage facility. The warehouse may be located at or near the firm or at some distant location, for example, near the market area being served. It may be owned by the firm or it may be an independent business operated for profit (an intermediary specialist).

Storage, as related to material movement and physical distribution, is a labor-intensive operation representing one of the more costly logistic activities. Cost aspects are enhanced by excessive handling requirements at each step of the process. Automation holds some hope for the future; however, the automated warehouse remains a myth rather than a reality.

Storage requirements exist in addition to those necessitated by material movement and physical distribution. These requirements are primarily concerned with internal needs of the firm and those resulting from a recognition of responsibilities extending beyond the point of ownership transfer.

The firm acquires internal storage needs when semifinished goods are removed from the transformation process for completion at a later date. This condition may arise because of a change in priorities, a delayed due date, a material shortage, or another reason. Additional storage requirements stem from the need to establish inventories of spare and repair parts to maintain the machinery and equipment used in the production process.

A storage requirement also arises when repairs are needed following product breakdown. The firm, or possibly an intermediary specialist, must acquire and store an inventory of spare and repair parts to assure their availability. The magnitude of inventory, hence the size of the storage facility, is a function of the

A. Quantity of the product distributed to the consuming public
B. Technological content of the product
C. Product reliability
D. Replacement time for additional spare and repair parts

The establishment of an inventory is a vital ingredient at each step of the storage process and entails a determination of (1) what should be included in inventory, (2) the quantity that should be included, and (3) when additional or replacement items should be ordered. Inventory requirements thus represent a major factor in determining the type and size of storage facilities.

Spares and Repair Parts

Spares and repair parts represent an investment of today's cash assets as insurance against a probable need in the future. To illustrate, assume a hypothetical work place illuminated by a single light bulb. The light bulb will fail at some time in the future; however, there is no way of determining just when it will fail. The expected life may exceed 2000 hours, but this is a statistical average. The actual life of any given light bulb may extend from a few seconds to a period well in excess of the expected life. To continue with this illustration, assume that a replacement bulb costs $7 and must be acquired at an outlet located several miles from the work place. The assumed minimum time to obtain a replacement is six hours, including travel to and from the "bulb store."

Under the defined assumptions, a failure of the light bulb will result in a minimum of six hours of no work if replacement items are not stored as spares. An expenditure of $7 of today's assets will, therefore, reduce lost time because of a failure to a matter of minutes rather than hours. Thus, in this example it is both cost-effective and efficient to expend today's assets as insurance against a probable future event.

Spares and repair parts, with the attendant development and maintenance of inventory, represent a critical element of integrated logistic support. Although normally associated with support following distribution of the product, spares and repair parts are equally important in support of the equipment used during the transformation process.

Personnel and Training

Training may well represent the most expensive element of integrated logistic support. Firms that produce complex, high technology, state-of-the-art products frequently require personnel with extensive training who are knowledgeable and proficient.

Training must be specifically designed and developed to be consistent with the product that is being produced, the applicable technical publications, the maintenance instructions, and the test and support equipment. Training must be scheduled to assure the availability of operators and repair technicians as needed to support the product (the right quantity of the right item in the right place at the right time). Careful personnel selection, coupled with an efficient training program specifically developed and integrated into a total logistics package serving the objectives of the firm, will minimize the cost of this logistic element.

Publications

A well-documented library of technical publications is essential to effective and efficient support of the product. Technical publications must be developed to properly serve their intended (anticipated) audience. As an example, the publication (and writing style for the publication) will be very different for support of a one-of-a-kind piece of equipment designed for the Research and Development Department of a major university than it would be for support of a relatively unsophisticated and commonplace household appliance.

Technical publications must include assembly and setup instructions, if appropriate to the product, and operating instructions. More elaborate documents may also include maintenance instructions, a listing of removable and replaceable parts, and the proper use of test and support equipment. The document must be consistent (integrated with the other logistic elements) and compared with the actual product to assure the accuracy and adequacy of the publication.

Test and Support Equipment

Machines and equipment, whether a part of the transformation process or a finished good distributed by the firm, require periodic adjustment, service, and repair. These activities may be greatly facilitated by the use of various types of test and support equipment. Test and support equipment should be specifically selected or designed to meet any unique requirements, the expected operating environment of the product, and the capabilities of operating and maintenance personnel. Allocated test and support equipment may turn out to be more complex than the equipment being supported, thus requiring additional logistic support.

Logistics involvement in test and support equipment includes a determination of what is required, the quantity required, and the schedule of availability (when the items will be required).

Facilities

The package of logistic resources remains incomplete until the facilities required to support the other logistic elements have been included. A logistic facility may require new construction or the modification of existing construction to meet specific logistic requirements. Logistic facilities include

A. A facility at the supplier for the storage of materials prior to shipment to the firm
B. A facility at the firm to store incoming materials until needed in the transformation process.
C. A facility for the storage of finished goods prior to distribution
D. A facility for the maintenance and repair of machines and equipment used during the transformation process
E. A facility for the maintenance and repair of the product following transfer of ownership

It is apparent that the facility requirements may be established through either the firm or intermediary specialists. Logistic activities include deciding the size of the facility, selecting the optimum location, and integrating the facility with other logistic elements.

SUMMARY

The initial approach to logistics was a horizontal array with a progression from subsistence logistics to operations logistics and systems logistics. The fallacy in this approach is that the subsets of logistics do not exist in isolation. Rather, the higher orders of logistics incorporate and build upon all lower orders.

Logistics in the context of the enterprise has been introduced as operations logistics and systems logistics, relegating the third subset, subsistence logistics, to a more primitive society. Operations logistics is concerned with the movement and storage of (1) materials into the firm, (2) semifinished goods through the firm, and (3) finished goods from the firm. Systems logistics, on the other hand, is dedicated to product support following ownership transfer. This separation of the logistics subsets is an artificial division of activities, since elements of both types are present in all phases of the industrial system. It is correct, however, to refer to the firm involved in the transformation of raw materials into finished goods as emphasizing operations logistics and to the firm providing product support following ownership transfer as emphasizing systems logistics.

Logistics is a support function in that it exists only to provide support to other functions of the firm. This support encompasses the logistic elements of transportation, storage, spares and repair parts, personnel and training, publications, test and support equipment, and facilities. The logistic elements do not reflect the total activity within the logistic arena, as many additional tasks are necessitated through implementation of a specific element. Examples of these derived tasks include a determination of the size and location of logistic facilities and the management and control of inventory.

The systems concept stresses a total integration of all functions of the firm (including the logistics function) toward the successful attainment of the objectives of the firm. Logistics and the integrated logistic support manager bear the responsibility of designing a coherent and harmonious package of logistic resources that optimizes progress toward the objectives. This frequently requires trade-offs and the suboptimization of functional review objectives for the good of the firm.

QUESTIONS FOR REVIEW

1. Do the elements of operations logistics also apply to subsistence logistics? Explain your answer.
2. Chapter 2 identifies a steel mill as a firm engaged primarily in operations logistics. Identify other types of firms that fit this description.
3. Name some types of firms that are primarily interested in systems logistics.
4. Do firms that use systems logistics place little or no emphasis on operations logistics? Explain your answer.
5. Explain the concept of reverse logistics.

6. Justify the use of a warehouse in providing support following the point of ownership transfer.
7. What is meant by the expression "a package of logistic resources characterized by harmony and coherence"?
8. Explain the concept of logistics as a support function.
9. Give an example of logistics support to the marketing function of the firm.
10. Give an example of conflict between two or more operating functions of the firm.

3

AN INTEGRATION OF LOGISTICS

Logistics is a vital ingredient of the firm, regardless of the firm's size. The sole proprietorship and the largest corporation depend equally upon an input of raw materials, a production (transformation) process, and distribution for the finished good. Inherent within these activities of the firm is a requirement for transportation (into and out of the firm) and storage (raw materials into the firm, semifinished goods within the firm and finished goods from the firm). Transportation and storage requirements in turn lead to a need for transportation and storage facilities and a method of managing and controlling these activities. The complexities of the production process may further require training in both the operation of equipment used during production and repair of production equipment. Operation and repair may in turn require equipment publications to document operation and repair procedures and the establishment of a parts inventory to facilitate repair. Repair procedures dictate a need for tools and test equipment.

A comparable allocation of logistic activities is required when support following transfer of ownership is included in the responsibilities of the firm. Consider, for example, a small corporation or sole proprietorship offering both sales and service of personal computers. It needs

1. A staff that is trained to permit an intelligent response to customer questions
2. A technical staff trained in the repair of previously sold units
3. A facility to enable repair of defective units

4. An inventory of repair parts (with attendant inventory storage requirements) to facilitate repair
5. A technical library, since it is impractical for service personnel to "remember" all operational, repair, and adjustment details
6. The transportation of both repair parts into the firm and defective or repaired units into and out of the firm
7. Tools and test equipment to effect timely repair of defective units.

The foregoing illustration reveals that logistic elements are not isolated from either the production process or from the support following the transfer of ownership. To the contrary, each logistic element may be appropriate to either or both activities of the firm.

Logistics Within the Firm

Integrated logistics within the firm is very much a potential rather than a reality. Individual logistic activities are frequently controlled and directed by various departments or functions within the enterprise. This diffusion of responsibilities not only increases the potential for duplication of effort and inefficiencies; it may also directly oppose the successful accomplishment of organizational goals. For example, consider a firm that utilizes private transportation resources owned by the firm and not a common carrier. A fleet of trucks controlled by a department within the production function may be required to transport the material for production into the firm. A fleet of trucks controlled by a department within the marketing function may be required for distribution of the finished product. A single control point within the logistics function of the firm may provide enormous benefit by coordinating transportation needs, whereby the vehicle distributing the finished product could transport a supply of raw materials into the firm on the return trip. Such a "back-haul" arrangement could provide a concurrent benefit by reducing total vehicular requirements of the firm by approximately one-half. This increase in efficiency of utilization and reduction in capital equipment expenditures has a positive impact in the profit potential of the organization.

The integration of logistics affords an equal promise of superior performance in both efficiency and profit potential when the logistics element inventory is considered. Inventory, as an activity of the firm, involves several diverse functions:

1. Raw materials for the production process must be stored for use as needed.
2. Semifinished goods must be stored until needed in the production process.
3. Finished goods must be stored until requested by the customer-consumer.
4. Spare and repair parts must be stored to repair production equipment.
5. Spare and repair parts must be stored to repair products that fail following the transfer of ownership.

Each of the functions listed above involve management and control of the acquisition and storage of their respective items for future use. The requirement for management and control of inventory is, however, independent of the reasons for establishing the inventory or its intended usage. The management and control process could very well be identical throughout the firm, yet the typical firm separates control of each inventory type by relegating managerial responsibility to managers of the various functions. The resulting potential for duplication of effort and inefficiencies can be dramatically reduced by delegating control to the integrated logistics manager.

Comparable analyses of the other logistic elements would reveal additional positive benefits and advantages of integrated logistics management.

Logistics: The Horizontal Function

The firm has a defined structure. The design of this structure is a major managerial problem, because it must permit employees to effectively and efficiently accomplish organizational goals. This structure allocates resources and people to tasks that are necessary in meeting the goals of the organization and provides for coordination within the firm.

The typical organization may be represented as a vertical structure having a well-defined delineation of responsibilities extending from the president or chief executive officer through various levels of management to the employee. Figure 3.1 illustrates a typical vertical structure for the production function of an organization.

A characteristic of the vertical structure is a well-defined and understood authority relationship. Employees A, for example, work for foreman A, who works for the manager of manufacturing, who works for the director of production, and so forth. Each employee receives direction and rewards (or punishment) from successively higher levels within the organization.

Logistics, however, represents an anomaly in that it does not follow the vertical structure so prevalent throughout industry. Figure 3.2 depicts the four functions of the firm (production, marketing, finance, and logistics) with the elements of logistics listed under the functions where they are most likely to occur. While all elements are identified under the production and marketing functions, it must be recognized that a given firm will employ only those elements appropriate to the industry. As illustrated in Figure 3.2, the logistic elements are divided between the production and marketing functions of the firm. Transportation is required during production, for example, to assure the acquisition and availability of materials needed during the production process. The marketing function also requires transportation to permit distribution of the finished good.

It is important to note that the logistics function does not include any of the identified logistic elements. Management of the individual logistic activities is under the direct control of other functions and departments within the firm. Yet the concept of integrated logistics requires that the ILS manager coordinate and control

Ch. 3 Logistics Within the Firm / 31

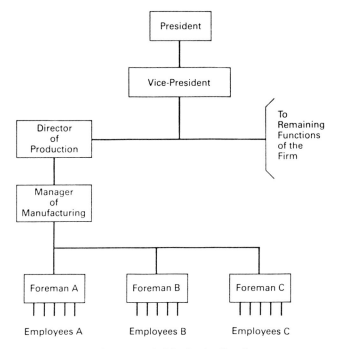

Fig. 3.1. A Vertical Production Function

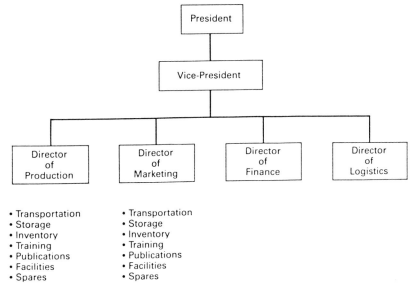

Fig. 3.2. Logistic Elements and Functions of the Firm

logistic elements to optimize success of the organization. This horizontal relationship, whereby one function of the firm (logistics) controls, coordinates, and manages logistical activities within other functions of the firm, represents a major challenge to the ILS manager.

A Potential for Conflict

Logistics within the organization frequently creates a very real potential for conflict. One source of this conflict is the logistic goal of one function of the organization, which, when considered in isolation, may be counter-productive to the goals of the firm. Consider, for example, an inventory of spare and repair parts that is maintained to facilitate the repair of equipment used during the production process. The manager of the group or section responsible for proper operation of production equipment wants a large inventory of spare parts, tools, and test equipment. This manager (and the personnel comprising the group or section) is evaluated on the status of production equipment and the relative speed with which malfunctioning equipment is returned to service. A large inventory greatly enhances the rapid restoration of operational capabilities.

This large inventory may, however, detract from the goals of the firm in that (1) the quantity of spare and repair parts, tools, and test equipment may require the acquisition and upkeep of additional facilities for storage; (2) management and control of the inventory may require additional employees; and (3) a large inventory represents operating capital that is tied up in nonliquid assets. Thus, it may be advantageous to the goals of the firm to accept an increase in the average time to restore malfunctioning equipment to allow a reduction in inventory requirements.

Conflict may also occur because of the potential for duplication and a lack of coordination within the logistic activity. As illustrated in Figure 3.2, logistic elements are distributed throughout the firm, with many of the activities duplicated in one or more of the operating functions. A previous example cited the benefits of combining the transportation element so that incoming materials and outgoing products could share a common transportation resource. This shared resource leads to conflict when the production and marketing functions have simultaneous and diverse requirements. Here again, logistics and the ILS manager must coordinate needs for the good of the firm. This coordination may reduce (suboptimize) the goals of one or more operating functions of the firm when these goals are considered in isolation from the overall goals of the firm.

The Logistics Function

Logistics, a support function, must complement and support the objectives of the firm. Any logistic activity that does not enhance profit, growth, or survival potential will cease to exist as a viable entity within that organization. Logistics does not, however, exist in isolation. It must function as a totality, in concert with the other functions of the firm and with the environment in which the firm operates.

Environmental Considerations

An organization cannot exist as a self-sufficient enterprise. It must reach beyond its boundaries for resources of production (personnel and materials) and a market for distribution of the finished good. The firm's growth and prosperity or loss of resources and expiration are inexorably linked to the environment in which it operates. In the broadest sense, the *environment* may be defined as those external (to the firm) factors that are capable of impacting the growth, profitability, or survivability of the firm. As illustrated in Figure 3.3, the environment may be separated into the general environment and the specific, or task, environment.

The *general environment* of the firm encompasses the cultural, economic, political, and sociological forces that influence the organization. These forces represent the characteristics under which organizations operate and are the same for all organizations within a given society.

The *task environment* exhibits a much more direct influence on the firm, and

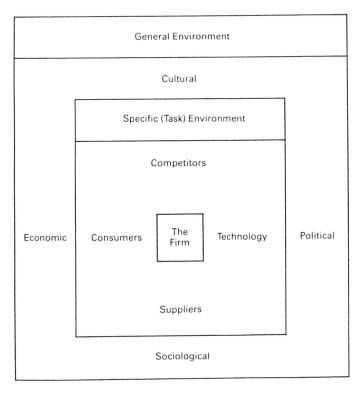

Fig. 3.3. The Environment of the Firm

successful interaction with these environmental considerations affords a distinct advantage over competing firms. Conversely, a failure to interact successfully may lead to failure of the firm as a viable business organization. The critical importance of the task environment was dramatically illustrated by the fuel crisis during the 1970s, which led to enormous increases in the cost of fuel. The transportation of materials into the firm and finished goods from the firm suddenly became a major cost factor. Added costs associated with this logistic element led to intense efforts directed toward increasing the efficiency of these logistic (transaction) channels.

Logistic Channels

The enterprise, regardless of size, must rely upon the environment for the resources of production and for the consumers of the products of production. A *logistic channel* is the process by which material flows into the firm and finished goods flow out of the firm and is that combination of resources needed for this movement. Logistic channel resources are rarely owned in total by the firm. Rather, the logistic channel normally consists of a number of independent enterprises acting together to deliver a product or material to the right place at the right time. The mechanics of logistic channels demand that each member perform assigned roles with a maximum of efficiency and a minimum of duplication.

A logistic channel consists of one or more enterprises operating as a system in providing a service to the firm. Figure 3.4 illustrates a logistic channel dedicated to providing material resources as an input to the firm. The logistic channel is made up of facilities and logistic specialists. A *facility* is an operating unit within the logistics process, and a *logistic specialist* is an independent business operated for profit. The latter is also referred to as an *intermediary specialist,* as it represents an intermediate point between the supplier and the firm (or between the firm and its customers). Successful operation of a logistic channel involves the activities of storage, transportation, handling, and communication.

In Figure 3.4, the logistical unit (warehouse) may be either. Warehouse 3 is owned by the firm for the purpose of storing incoming materials until needed in the production process. This warehouse is a facility. Warehouse 2, on the other hand, is a private enterprise, thus it is a logistic specialist. Warehouse 1 is used to store the output (finished good) of supplier 1 until a request for this material has been received from the firm. The request for material is a part of the communications function. Upon receipt of the request, this material is loaded into the transport vehicle (the handling function) for movement to the firm. The material is now loaded into warehouse 3 until needed in the production process. Warehouse 2 is included in Figure 3.4 to illustrate an alternate path in the logistic channel. Warehouse 2 may accumulate a variety of materials from various suppliers. The resulting material assortment is stored pending receipt of a material order from one or more firms.

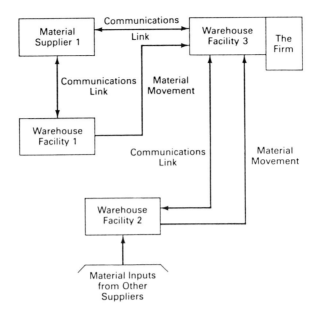

Warehouse 1: Supplier Owned
Warehouse 2: Private Enterprise
Warehouse 3: Owned by the Firm

Fig. 3.4. The Logistic Channel

This analysis works equally well for product distribution. Consider, for example, the food industry, where the output of farms, canneries, and food-processing firms is stored in a regional warehouse, thus creating an assortment of goods. Supermarkets in the area served by this warehouse may request a variety of products from a single source (the warehouse). This provides a significant advantage over the alternative of placing separate orders with individual suppliers.

Logistic Operations

Logistic operations includes all activities involved in the successful completion of a logistic activity. Logistic activity is, in turn, initiated with the need for logistic support and a corresponding request for logistic service. Logistic service may involve a raw material, the finished good, a training course, needed repair action, or any other logistic activity. The logistic operation is concluded with an acceptable response to the request for service. Thus, it can be concluded that logistic operations incorporate a performance cycle that begins with the identification of a logistic need and concludes with an acceptable resolution of that need.

The Performance Cycle

Logistics is interconnected with other functions of the firm, material suppliers to the firm, and customers of the firm. This interconnection involves transportation and the associated communications that stimulate the need for a logistic operation, and incorporates a material cycle, an inventory transfer cycle, and a product cycle into an overall logistics performance cycle. Figure 3.5 illustrates the logistic cycles of activity that comprise the logistics performance cycle.

A logistics performance cycle is initiated upon recognition of a need for logistic service. Awareness of this need leads to a request for service (the communications link). Requests may originate with the customer or consumer (the product cycle) or with the enterprise (the material cycle or the internal inventory transfer cycle). The communications link, in turn, triggers the logistic activities necessary for the movement of material into the firm, semifinished goods through the firm, or finished goods out of the firm.

Figure 3.5 incorporates material storage and product warehousing facilities as a part of the enterprise. Few products are produced to order, thus specific amounts of material are stored by the firm in anticipation of future need during the production process. Following this same line of reasoning, a limited variety of the finished good (products of the firm) is also stored in anticipation of future demand. It is important to note that the material and product warehousing functions may also be

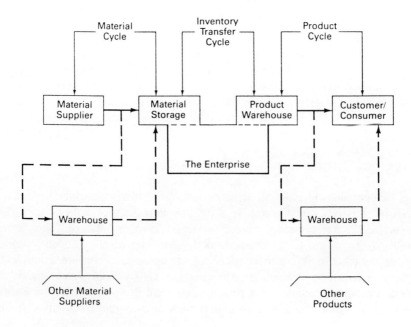

(– – –) indicates an alternate channel.

Fig. 3.5. The Logistic Performance Cycle

fulfilled by logistic specialists (intermediary specialists), who provide this service to the enterprise for profit.

The alternate path (broken line) on the material side of the logistics performance cycle triad in Figure 3.5 illustrates a firm that requires a diverse assortment of material from various suppliers. This multiplicity of suppliers may create an assortment of goods at a centrally located warehouse. This warehouse then becomes the material source for the firm. Similar logic prevails when considering an alternate path within the product cycle. The customer may purchase a diverse assortment of goods from a common warehouse facility that acquires those goods from a variety of firms.

Logistics performance cycles are measured by their effectiveness and efficiency. *Effectiveness* defines how well objectives are met; *efficiency* is related to the expenditure of resources in meeting those objectives. As an example, consider the highly unlikely objective of providing the best possible degree of service (for dishwashers, for instance) following the point of ownership transfer. The ultimate in service would be to have a service technician available for each dishwasher that is sold. This would be very effective in meeting the objective of providing the best possible service; however, it would certainly not be efficient when measured against the expenditure of resources.

It should be apparent that the customer is endowed with ultimate control over the logistics performance cycle. The request for a product by the customer begins the product cycle, with the subsequent purchase representing a decrease in product inventory at the firm. This decrease in inventory stimulates the production process, and material begins to flow through the firm (the internal inventory cycle) as raw materials are transformed into the finished good. The production process consumes the materials of production, thus triggering a request for additional material from suppliers to the firm (the material cycle). This control by the customer reflects a philosophy referred to as the *marketing concept.*

The Marketing Concept

The change to a market-based economy wherein the buyer is "king" is a relatively recent innovation. Prior to the 1950s, goods were scarce and alternate selections were few. The goal of the enterprise was to produce, and the customer was given little choice beyond the option to buy or not to buy. This production-oriented philosophy focused almost entirely upon the ability of the firm to produce and sell the product. The customer was considered as little more than a receptacle for the goods of production.

The increasing availability of consumer goods and the growing competition resulting from alternative sources led to the emergence of a market orientation during the 1950s. The shift to a marketing concept represented more than a shift of emphasis. It was a new philosophy of business, one that required the firm to survey the marketplace to determine what was needed and what was desired and to produce products that satisfied those needs or desires. "Give the customers what they want" became the new business creed.

The market-oriented methodology, typical of today's market, represents several new and different challenges to the manager. It (1) requires the manager to know the markets of the firm and where they are located; (2) introduces the concept of effective customer service and produce support; (3) places the right product in the right place at the right time and at the right price; (4) establishes the need for distribution channels to maximize sales at the right price; and (5) requires adequate support of the product following ownership transfer. Notice the close parallel between the marketing concept and the concept of integrated logistics support. It is no mere coincidence that the rise in logistics has paralleled the growth of the marketing concept.

Since the customer is now a dominant force, market opportunities must be studied to assure survival of the firm. The enterprise must now determine what products are needed and are likely to be purchased. A product that is economical to produce generates little profit if it is not needed or desired by the consumer.

Products, whether decided by the firm (a production orientation) or by the consumer (a market orientation), determine the distribution and allocation of resources within the enterprise. The transformation process, through utilization of these resources, produces finished goods that must be transferred to the customer. Long-term production of goods can only be maintained through a continuation of the ownership transfer process. The firm can no longer continue as a viable enterprise when the goods of production remain unsold.

A market-oriented firm uses market surveys to ascertain which products possess a high potential of being needed or desired by the consumer. Logistics, as a critical function of the firm, plays a key role in the process by which a prospective customer gains access to, accepts, and eventually trusts a new product. This role varies with the stages of the product life cycle.

The Product Life Cycle

Products, as well as people, follow a natural progression from birth, through growth and maturity, to retirement and death. Product progression, which may vary from a few months to several decades or longer, is referred to as the *product life cycle*. The four phases of the product life cycle are illustrated in Figure 3.6 and are the Product Introduction phase, the Market Growth phase, the Market Maturity phase, and the Product Decline or Phase-Out phase.

New products being life at the *Product Introduction* phase, which is characterized by limited market acceptance. The new product must be heavily promoted, since the consuming public possesses, at best, a limited awareness of the existence, advantages, and uses of the item. Logistics, during this introduction phase, is primarily concerned with the establishment of highly efficient channels of distribution. Transportation assets capable of providing the materials of production and assuring product availability in response to customer demand must be developed.

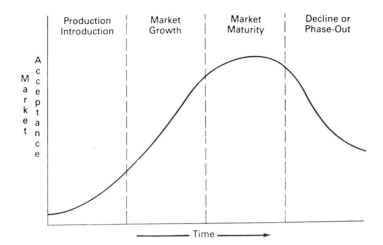

Fig. 3.6. The Product Life Cycle

Warehouse facilities must be provided for storage of production materials and finished goods. An inventory management system must be set up to control incoming materials, semifinished goods, and a limited inventory of the completed product. Product complexity and expectations of increasing consumer acceptance may necessitate setting up a customer service network. Logistics costs per unit are relatively high because of (1) costs of the logistics resource setup, (2) required high levels of product availability, and (3) the necessity for a rapid response to customer orders.

The product that gains acceptance by the consuming public enters the *Market Growth* phase. During this phase, sales are rapidly increasing and profits are high, as competing products have yet to enter the marketplace. This second stage may last from several days to several years. Logistics performance during the Market Growth phase shifts from the previous emphasis on high performance to a more realistic balance between cost and service. Limited competition and high demand permit a wide latitude in managing product distribution channels in order to minimize the logistic costs per unit.

The product that has gained market acceptance and is enjoying increasing sales attracts more and more competitors wishing to enter the race for profits. This period, characterized by aggressive competition, identifies the *Market Maturity* phase. Intense competition leads to declining profits with increased demands for efficiency in the use of logistic resources. The logistic emphasis again shifts to high levels of availability and service as the firm attempts to retain and add to its customer base. The average logistic cost per unit increases with the increased attention accorded customer service.

The *Product Decline* or *Phase-Out* phase is characterized by a shrinking market. The product is no longer in demand. Logistics must continue to support the product while avoiding unnecessary risk should management elect to cease production. Minimum risk has now become the driving force within the logistics arena. Logistic support, however, may not end with the decision to discontinue production. The firm may very well have incurred an obligation to provide product support for a stated period of time following production. In this instance, logistics must assure that spare parts and customer service facilities continue to be available.

Coordinating the Logistics Function

Logistic operations is defined as the activity that takes place when a function within the firm is tasked with the accomplishment of a logistics objective. The logistics objective may take the form of

- A. The acquisition and storage of the materials of production
- B. The storage and physical distribution of the finished good
- C. The acquisition, management, and control of an inventory to support the equipment of production or the finished good
- D. The establishment of logistic channels
- E. The development of training courses to support production equipment or as a customer service
- F. The preparation of documentation necessary to support operation and maintenance of the product

or any of the other diverse objectives inherent to logistics.

The activity of logistics must, however, be preceded by the need for this activity. When should additional material for production be requested? When should a training course be scheduled or the service facility placed in operation? Logistic activity must occur in response to a need, and the need must be coordinated with the activity. Coordination is the activity that integrates the requirements for logistics and logistic operations into a coherent and harmonious whole.

The customer, by acceptance of the finished good, is the ultimate determinant of the need for logistics. The product that remains unsold requires no physical distribution and negates the need for additional materials. From this it is apparent that a determination of product demand is critical to the logistics function. Product demand leads to sales, which deplete the inventory of finished goods, thus creating a need for the production of additional goods. This, in turn, creates a need for the materials of production. Material needs lead sales by an amount of time approximating the production process plus logistic channel times. Therefore, the inventory must be closely managed to prevent a surplus or a shortage. This leads to the inescapable conclusion that the key factors in coordinating logistic needs with logistic operations are (1) the prediction of future demand through market forecasting and (2) material management.

Forecasting

Estimates of market potential and probable sales volumes are vital to effective coordination of logistic elements. Market opportunities that may be available must be evaluated through forecasting to permit physical distribution planning and establishment of the need, or lack thereof, for customer service facilities.

Market forecasting is an attempt to reduce the impact of demand uncertainty through product sales projections. A variety of techniques are used in forecasting; however, they fall in one of two basic categories: (1) an extension of past performance and (2) an anticipation of future performance.

Forecasting of existing products usually implies the existence of historical data that permit a projection of past sales figures into the future. The basic approach is a simple extension of past sales trends into the future (see Figure 3.7). The fallacy of this trend-extension method is that it assumes conditions in the past will continue into the future. In fact, the future does not replicate the past, and this projection will be wrong whenever future conditions lead to a fluctuation in demand. For this reason, forecasters seek other techniques to aid in anticipating future events.

Anticipation of future performance is an attempt to predict what *will* happen, rather than simply extending what *has* happened into the future. Some of the methods used are (1) opinions of executives within industry, (2) estimates by sales personnel, (3) surveys of the marketplace, and (4) market testing.

Forecasting market demand for new products is the most difficult and risky task of all. The new product has no relevant historical data that can be projected, and it is unrealistic to expect sales personnel or the consuming public to have valid opinions about an unfamiliar product. All is not hopeless, however, as an analysis

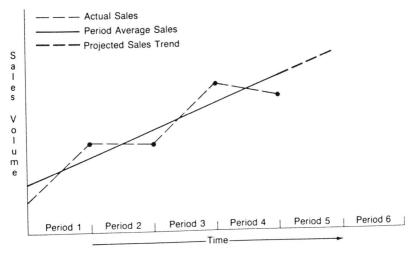

Fig. 3.7. Linear Trend Projection

of products that the new item may displace can provide an indication of potential demand. The potential demand yielded by this substitution method may then be scaled down by market realities and sound judgment.

Regardless of the method used, the purpose of forecasting is to generate expectations concerning future demand. Analysis of this demand may then be used to evaluate the demands placed upon the logistics function, e.g., what quantity of the product should be distributed and where and when it should be distributed.

Material Management

Product demand stimulates the logistic activities of physical distribution and related product and customer support functions. Physical distribution, in turn, reduces product inventory, thus creating the need for additional production. Production requires the movement of both semifinished goods through the firm (internal inventory transfer) and the materials of production into the firm. Coordination and control of these are the function of material management. Figure 3.8 illustrates a simplistic example of the material management process.

With reference to Figure 3.8, the product demand is projected as 2 units in period 1, 2 units in period 2, 4 units in period 4, and so forth. The on-hand inventory (material available for production of the product) is identified as 4 units. Assuming that 1 unit of material is required to produce 1 unit of the product, the product demand during period 1 (2 units) will consume 2 units of material. This reduces on-hand inventory to 2 units. The 2 units of on-hand inventory are consumed in period 2 in order to meet the expected period 2 demand of 2 units. This reduces internal inventory to zero.

Period 4 identifies a demand for 4 units; however, the internal inventory is at zero. In order to produce these 4 units of the product, 4 units of material must be

Period	1	2	3	4	5	6	7	8	
Demand	2	2	0	4	2	0	2	2	Product
Expected Receipts				4	2		2	2	Material
On-Hand Inventory 4	2	0	0	0	0	0	0	0	Material
Order Requests		4	2		2	2			Material

Lead Time: Two Periods

Fig. 3.8. The Material Management Process

received no later than this period. The identified lead time of two periods indicates that a material order placed in period 2 would be expected to arrive at the firm in period 4, thus permitting production of the 4 product units.

This example would work exceedingly well if we could assume that expected demand is equal to actual demand. This, however, is not the case. Actual demand fluctuates in response to numerous unforeseen factors, and any estimate of future performance is going to be in error. Errors may be due to demand in excess of projections (type 1 uncertainty) or to increases in the lead time because of delays in the material acquisition cycle (type 2 uncertainty). Safety stock may be retained in inventory to protect against either type 1 or type 2 uncertainty.

It is apparent that errors will also occur because of a decrease in demand or in the material lead time. This error is not normally a problem, as either instance would lead to an increase in the on-hand inventory. This could become a problem, however, if (1) the decrease in demand continued to a point where sales could not support production or (2) the decrease in lead time resulted in excessive material storage requirements.

A decision to protect against type 1 and type 2 uncertainty by establishing safety stock must be evaluated in terms of the associated inventory cost and an acceptable degree of risk. With reference to Figure 3.8, assume that market research and statistical analysis indicate that actual demand is within 2 units of estimated demand 95 percent of the time (approximately 2 standard deviations). Actual demand during period 4 may then be expected to be at least 2 units but not more than 6 units with 95 percent certainty. If the resulting 5 percent risk of having to stop production because of a shortage of material (stock-out) is acceptable, a safety stock of 2 units should be established, and the on-hand inventory should not be permitted to decrease below that level. Figure 3.9 illustrates the application of safety stock to the material management process.

The increased use of computers in the firm and the growing sophistication of forecasting techniques have contributed to the growth of material management as a

Period	1	2	3	4	5	6	7	8
Demand	2	0	2	4	2	0	2	2
Expected Receipts				4	2		2	2
On-Hand Inventory 6	4	4	2	2	2	2	2	2
Order Requests		4	2		2	2		

Lead Time: Two Periods
Safety Stock: Two Units

Fig. 3.9. An Application of Safety Stock

profession. Additional advances may be expected to reap substantial dividends as inventory requirements are further reduced, thus freeing capital for alternative uses.

SUMMARY

Logistics is a critical element of the enterprise because it depends upon raw materials, the transformation process, and distribution of the finished good. A more recent addition to the logistics arena is providing support beyond the point of ownership transfer. Yet the integration of logistics in corporate America remains very much a potential rather than a reality. A major factor contributing to this lack of integration is the distribution of logistic elements throughout the functions of the firm. Logistics is a horizontal arrangement, unlike the more conventional vertical organizational structure. This structure leads to a high potential for conflict between the functions of the firm.

The firm, and by extension the logistics function of the firm, must exist within the constraints of the environment in which it operates. The firm looks to the environment for resources that enable production and for customers to consume the products of production. The vehicle for connecting the firm to its environment is the logistic channel. Logistic channels provide for the transportation of material into the firm and for the physical distribution of completed products.

Logistic operations refers to the activities that are performed to attain an objective. Logistic operations, through cycles of performance within the logistic channel, connect the firm to its supplier (the material cycle) and the firm to its customers (the product cycle). A third cycle, the inventory transfer cycle, provides for the movement of semifinished goods through the firm. The three cycles make up the logistics performance cycle.

The 1950s witnessed the beginnings of a change from a production-orientation to a market-oriented philosophy, where the consumer is the dominant force in the economy. The firm could no longer dictate to the customer; it must now look to the customer to determine the products that are needed and desired. Market research and forecasting can provide some indication of customer preference, however, these are far from infallible. Logistics now faces the added challenge of an emphasis on customer service and demands for increases in the efficiency of logistic operations.

Forecasting provides the information necessary to project an estimate of future demand. The magnitude of this demand determines how much of the product should be produced. This, in turn, provides information on the amount of material that must be acquired to sustain production. Estimated product demand permits logistics to evaluate the required physical distribution resources and the need for customer service facilities. Product demand projections also provide an indication of the need for the materials of production. Therefore, logistics can evaluate material cycle requirements. The knowledge of what, when, and how many products and materials are needed by the firm permits coordination of logistic operations.

QUESTIONS FOR REVIEW

1. Why is logistics defined as a horizontal function?
2. Explain the concept of logistics as a support function.
3. Define the task environment of the firm.
4. What is the logistic channel, and how does it relate to the logistics performance cycle?
5. What are the advantages to the firm of a warehouse that has an assortment of goods from various suppliers?
6. What are the characteristics of a market-oriented firm?
7. List several products and place them on the proper phase of the product life cycle.
8. Explain the need for a highly efficient logistics function during the Introduction phase of the product life cycle.
9. Why can we say that the customer is the ultimate determinant of the need for logistics?
10. How does safety stock protect against type 1 and type 2 uncertainty?

4

LOGISTIC PREREQUISITES

Logistics may be considered in the limited context of support to the firm (operations logistics) or in the broader context of support to the product (systems logistics). It is well to remember, however, that the division into operations logistics and systems logistics is largely artificial. All logistic elements are present in both subsets, although there are recognizable differences in emphasis between the two. Operations logistics, for example, is primarily concerned with the movement of material into the firm (material management), the movement of semifinished goods within the firm, and the movement of finished goods from the firm (physical distribution).

Implicit within these requirements is the need for storage facilities (warehouses) for storage and control of the inventory of (1) the materials of production, (2) semifinished goods, and (3) the finished good. The remaining logistic elements (training, spare and repair parts, technical publications, and tools and test equipment) are relegated to supporting the equipment of production.

Systems logistics, with an emphasis on support of the product, is primarily concerned with physical distribution of the product to customers and customer support following ownership transfer. This latter requirement entails the establishment of parts inventories, repair facilities, technical publications (related to the product), and personnel training to develop the skills and knowledge needed in providing this support.

This shift in emphasis is of little consequence to the logistic practitioner. Training, for example, follows the same basic procedure in meeting the objective of supporting the equipment of production as it does in supporting the products of production.

The ILS manager, in either instance, orchestrates the elements of logistics to assure having the right quantity of the right item in the right place at the right time.

The objective of harmony and coherence among logistic elements cannot be attained solely through the efforts of the ILS manager. Successful achievement of logistic objectives is, to a significant degree, predetermined during the product design and development process. The $100 million item of production equipment and the $100 consumer product must each be designed for logistic supportability. An item or product that cannot be repaired eliminates the need for repair facilities, a parts inventory, or training in repair procedures. It can only be replaced. Thus the logistics of customer service is limited to little more than transport of the replacement item or product.

A Balanced Integration

The objective of integrated logistic support is quite simply to develop a package of logistic resources that optimizes the operation of any system, whether military, industrial, or consumer. The concepts and practices of ILS apply equally to the military fighter plane, the automobile plant, and to toasters. There is a difference, however, in the application of logistics to existing products as opposed to its application to new products or to existing products that have undergone a major change.

Existing products require logistic support, but the type and quality of this support are largely determined by product features and characteristics. The automobile or air conditioning system that is in production today is a product of yesterday's design and development. Logistics must work within the constraints that exist as, for example, the product that is extremely difficult to repair cannot be changed in the short term. Logistics can, at best, strive for increased efficiency of performance. It is incapable of rendering the product easier to repair. Such changes can only be effected by introducing new products or extensive design changes in the existing product.

The new product begins with an idea, then evolves through concept exploration and feasibility development models into the final product. Logistics, to be effective, must enter the product development cycle near the beginning of the concept exploration phase. Characteristics of the product that enhance logistic supportability must be designed into the product; they cannot be added on after the fact. Such characteristics are required for an optimum logistics program and are derived from several design-related activities that may be viewed as prerequisites to logistics.

Prerequisite activities offering significant contributions to logistic supportability include reliability, maintainability, repair level analysis, and life cycle costing.

Reliability is related to the need for repair actions. The product will fail as infinite reliability exceeds practicality. The question therefore becomes one of how often the item will fail. The design of an optimum support package is partially based on the anticipated frequency of failures, and reliability may be considered the starting point for integrated logistic support. *Maintainability* is the ease of repair (or lack thereof) of an item. Knowing that things will fail, the logistic supportability problem may be simplified by designing a product that is easy to repair when it does fail. *Repair level analysis* provides a rational approach to the problem of what should be repaired and where is the most efficient location for repair. *Life cycle costing* reduces the entire life span of the product to the common denominator of dollars and cents.

The activities of reliability, maintainability, repair level analysis, and life cycle costing are the duties of design engineering and logistics. They provide the connecting link between design of the product and design of the logistic support package. The characteristics of logistic supportability cannot be added onto the finished good; they are included with each step of the design and development process. For this reason, the logistician must be included as part of the team leading to the development of each new product. This is equally true for major changes to existing products, as changes to existing products may be considered the same as new products, in that logistic practitioners have similar opportunities to help increase logistic supportability.

A balanced integration of logistic resources and the product can only be achieved through application of these design-related specialties.

Reliability

Design begins with a concept; it continues through engineering design and development; and, assuming feasibility of the original concept, culminates in the new product. Exacting engineering disciplines have properly arranged the materials of production to create a new product capable of performing its intended function. The product will perform, but will it continue to perform over time? Will the product provide long-term satisfactory service, or will it be inoperative much of the time due to frequent malfunctions? In short, is it reliable?

Reliability, if it is to exist as an inherent characteristic of the product, must be built into the product by reliability engineering during the design and development process.

Elements of Reliability

Reliability is nothing more than the probability that a product will perform as expected over a defined period of time and under specified conditions. Implicit within this definition are the four elements that determine the reliability of a product—probability, performance as expected, time, and specified conditions.

The first element, *probability,* is stated as a quantitative expression representing the percentage expectation of satisfactory performance. There is, for example, a 99 percent probability that an automobile will start tomorrow morning and provide a means of transportation to the work place (a distance of 5 miles through heavy traffic). This indicates that one can expect the automobile to function as expected (complete the drive to work) 99 times out of 100. There is a fallacy in this simplistic example, however, as probability applies to the total population of the item being measured (the automobile in this example). A given unit may fail to provide satisfactory service three-fourths of the time (25 percent reliability), whereas another unit may greatly exceed expectations. The average of all failures, for all units being considered, over the defined range of performance provide a reliability figure for that item. Thus, it is apparent that reliability is based upon the mathematics of probability.

Performance as expected is the second element of reliability. This implies that specific criteria must be established to define the expected performance as accurately as possible. Again returning to the example of the automobile, a car with no brakes may be capable of starting and, with a great deal of luck, be capable of providing transportation. Should this vehicle be considered as providing the expected performance? Of course not! Expectations regarding performance are normally based against product specifications, and automobile specifications include an ability to stop as well as to go forward and reverse. Satisfactory performance includes both quantitative and qualitative factors, however. Certain drivers, for example, may consider an automobile with a defective muffler as acceptable, whereas others would consider it out of service.

The third element, *time,* is one of the most important, as it provides a finite factor against which product performance can be measured. The term *time* is somewhat of a misnomer, however, as operating hours represent only one method of measurement. Product performance may also be measured in miles of operation, number of events, number of cycles, or any other convenient standard of repetitive units, for example, a camera, that assumes 1 million operations (cycles) of the shutter mechanism without a failure. Time, in conjunction with the other reliability elements, provides all data necessary for predicting failures within a sustained period of operation. For example, a backup power generator for the home may have a rating of 1.2 failures for each 2,000-hour period of continuous operation. This indicates that 11 failures may be expected during a 10,000-hour operating period. This measurement leads to *mean time between failure* (MTBF), a common definition of reliability. MTBF for the generator in this example may be determined by dividing the stated operating interval (2,000 hours) by the number of expected failures during that interval (1.2).

$$MTBF = \frac{2,000 \text{ hours}}{1.2 \text{ failures}} = 1,667 \text{ hours per failure} \quad (4.1)$$

Equation (4.1) indicates that one failure may be expected for each 1,667 hours of operation. Conversely, if the generator has been operating for 1,100 hours, a failure may be expected within the next 567 hours of operation.

Specified conditions make up the fourth element that determines product reliability. These conditions define the operational profile and include both environmental factors and the physical attributes (vibration, shock, and so forth) of the operating location. Specified conditions are not limited to the operating scenario, however, as the handling of the product during transportation and storage are equally important. What, for instance, would be the reliability of a personal computer (unprotected by any packaging) transported over rough roads in the back of a truck and stored out of doors until requested by the customer?

Tools of Reliability

Reliability must be designed into a product; it is an inherent characteristic of design and cannot be added on after the fact. The tools of reliability include (1) redundancy, (2) simplification of design, (3) Hi Rel parts, (4) derating of parts, and (5) environmental constraints. Hi Rel stands for high reliability, obtained through the application of stringent quality control during the parts production process. Derating, on the other hand, refers to the use of parts with a higher than required rating.

Including reliability into the design of a product entails the use of additional resources during the design and development phase. The added expenditures represent an investment in the future, and increased reliability translates directly into decreased failures over the life of the product. Decreased failures, in turn, lead to reduced expenditures for extensive (and expensive) maintenance and repair actions. An additional benefit of increased reliability is evident in a consumer-oriented society. Repetitive and frequent failures may result in a loss of the firms' customer base as consumers search for competitive (and more reliable) products. Increased reliability is expensive and redundancy is an apt illustration of the added expense which may be encountered when steps are taken to increase product reliability. First, consider the series relationship that represents the most widely used type of product. The series relationship, illustrated in Figure 4.1, incorporates a serial sequence of components or parts wherein each part must operate if the product is to function correctly. As shown in Figure 4.1, product reliability is equal to the product of the reliability of each part, and the failure of any one part equals the failure of the product. The product is therefore only as reliable as the "weakest" part in the chain.

The reliability of a product characterized by a series of parts may be improved by using parts in parallel as illustrated in Figure 4.2. In this relationship, identical

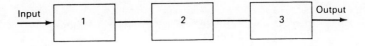

Reliability of a Series Network: $R_{SN} = (R_1)(R_2)(R_3) \cdots (R_n)$

Fig. 4.1. A Series Relationship

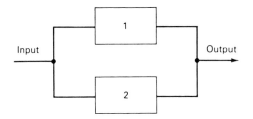

Reliability of a Parallel Network: $R_{PN} = R_1 + R_2 - (R_1 R_2)$

Fig. 4.2. A Parallel Relationship

parts are in parallel and both parts must fail before the product fails. Note that even though the reliability of individual parts may be exactly the same as for the series relationship, the reliability of the product has increased as the failure of a single part does not result in product failure. This increase in reliability has been gained, however, at the added expense of duplicate parts, an expense that may be further increased through added weight, a more complex design, and increased space requirements.

Does the increase in reliability warrant the added expense? That is a question that can only be answered after an evaluation of the cost, the probability of failure, and the likely consequence of failure.

Redundancy and the other tools (simplification, Hi Rel Parts, Parts Derating and environmental constraints) may each be used to meet the objective of increased reliability. It is expensive, but the cost of an unreliable product is frequently many times more expensive.

A Failure Rate Curve

The reliability of a product is also measured by the failure rate, which is determined by dividing the number of failures by the total operating hours. For example, a product that experienced five failures during 1,000 hours of operation would have a failure rate of 0.005. This is nothing more than the reciprocal of the MTBF, which is 200 hours for a product having five failures in 1,000 hours of operation.

The failure rate (reliability) curve is not uniform over the life of a product. A relative evaluation would reveal that a high initial failure rate is followed by an extended time interval characterized by a much lower rate. This is, in turn, followed by an increased rate of failure as the product nears the end of its useful life. The "bathtub curve," a name descriptive of its shape, is illustrated in Figure 4.3.

The failure rate (bathtub) curve covers all failures experienced by a product. Failures occurring early in the operational life of a product, for example, the many repairs and adjustments frequently needed on a new vehicle, are called *infant mortality failures*. Infant mortality is decreased through testing, burn-in, debugging, and other activities designed to assure that products are operationally ready.

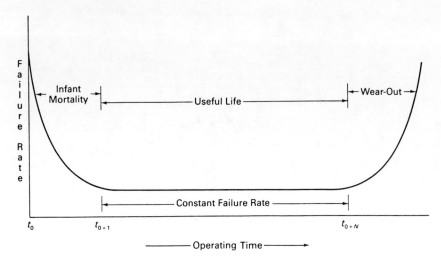

Fig. 4.3. A Failure Rate Curve

Burn-in refers to the process of monitoring product operation over a period of time prior to sale of the product to the customer or consumer.

Infant mortality is followed by the useful life span and is, hopefully, much longer than the other two phases. It is assumed, for statistical simplicity, that the failure rate is constant throughout this period and failures that occur are randomly distributed. The reliability engineer's task is to bring this portion of the curve as close to zero as possible.

Increased failure rates near the end of the curve signify the wear-out phase. This portion of the curve applies to that part of a product's life span when mechanical parts and other components are aging.

Maintainability

A product is going to fail! Reliability engineering may increase the time between failures (decrease the failure rate), but failure remains the inevitable result of continued operation. Failure, with the exception of specific throw-away items, is normally followed by the restoration of operable status through maintenance and repair.

The product that fails represents an expense in that it is not performing its intended function. For example, the piece of production equipment that fails may halt an entire assembly line, thereby forcing dozens of workers to be idle, or an automobile may fail, thus preventing the worker from reporting to work, which reduces his or her personal income. This expense is further increased through the high cost of maintenance required to restore the item to service.

The inevitability of failure and the costs of service suggest that ease of maintenance, or maintainability, should be a critical product objective.

Maintainability is the probability that a product can be restored to operational use within a specified period of time. It is an inherent characteristic of design, and the maintainability engineer must influence the design process if the product is to be easily maintained.

Units of Measure

Maintainability engineers must be design specialists as well as technicians. Competence in designing is needed to assist the product designer in creating a maintainable product, and a technician's viewpoint is required to facilitate product maintenance. The design specialist must determine the objectives of the design, quantify the maintainability elements, and decide how maintainability is measured.

A frequently used measure of maintainability is maintenance man hours divided by operating hours (MMH/OH), which defines the total maintenance expended over a finite interval of time. The maintainability measure, MMH/OH, by defining maintenance expenditures as a function of time, indicates the number of maintenance personnel required. This method of measuring maintainability, however, places the maintainability engineer at a distinct disadvantage, since MMH/OH is a function of maintenance time as well as maintenance frequency. Maintenance frequency is derived from MTBF, which is a reliability measure. Thus, if the reliability engineer fails to meet MTBF objectives, the maintainability engineer is almost certain to miss MMH/OH goals.

A second maintainability unit of measure is *mean downtime,* which is the statistical mean of the time a product is out of service for maintenance. This measurement indicates inherent product availability. For example, the automobile that accumulates an aggregate out-of-service time of two weeks in a period of one year is available for use during the remaining fifty weeks. The availability of this product is just over 96 percent (fifty weeks available for use divided by fifty-two weeks of potential availability).

Mean time to repair (MTTR) is yet another measure of maintainability. The product that is easy to maintain can quickly be restored to service; it exhibits a preferred MTTR. This unit of measure must not be used by itself. A product having an excellent MTTR (that is, one that can be very quickly repaired) exhibits a questionable virtue if it also possesses a very low MTBF (that is, has a high failure rate). Net maintenance time should also be included in the calculations deriving the product MTTR.

Net Maintenance Time

Units that measure maintainability rely on the time required for maintenance. In fairness to the product and the responsible engineers, it is important that this figure accurately reflect true repair time (net maintenance time as opposed to gross maintenance time).

Consider the following example illustrating the difference between net and gross maintenance time. An automobile has a flat tire. The tire can be removed and replaced in twenty-one minutes; this is the net maintenance time. However, the total time charged to this maintenance task may reflect an expenditure of several hours. The difference lies in the multitude of associated tasks that must be performed in order to accomplish the primary objective of replacing the defective tire. These associated tasks may include the following:

1. Get keys to service vehicle.
2. Pick up assistant.
3. Go to tire shop.
4. Drive to disabled vehicle.
5. Take defective tire to tire shop.
6. Complete paper work.
7. Park service vehicle.
8. Return keys.

The time required for all associated tasks may be several hours. This time, plus the twenty-one minutes of net maintenance time (time actually required to replace the defective tire) is the gross maintenance time. The associated tasks were necessitated by the failure, therefore it is correct to allocate this time to the repair activity. Associated tasks cannot be affected in any way by the design of the product, so it would be incorrect to penalize the product (and the maintainability engineer) by using gross maintenance time in maintainability calculations. This aspect of the task is beyond control of the design process.

Repair-Level Analysis

Reliability measurements indicate how often a product will fail, and maintainability measurements provide information on how easy (or difficult) the product is to repair. Next is the decision where the repair action, if any, should take place. Should the product be repaired or simply discarded? If repair is warranted, should the product be repaired at the location where it is being used? Should it be returned to the manufacturer's repair facility or to a third party (intermediary specialist) offering this service? Should it be returned to a centrally located facility that specializes in the detailed repair or rebuilding of the product? Repair decisions such as these must be made for the product and for each removable component of the product.

Repair versus Discard

A necessary prerequisite to determining where repair is to take place is to determine whether the product or item should be repaired. For example, consider the electric light bulb. Today's technology certainly possesses the capability of restoring a burned-out light bulb to useful life. The cost of this repair, however, would far exceed the expense of simply discarding the defective item and acquiring a replacement unit. The intelligent and cost-effective decision is to throw away the defective light bulb; repair is not economically feasible. On the other hand, consider a more complex product such as the personal computer. Here the cost of repair is (hopefully) much less than the cost of a new unit, thus repair is the more attractive choice.

Repair-versus-discard decisions exist on two different planes. First, the new product and its parts and components must be evaluated concerning the economic feasibility and practicality of repair. Second, the product that is in use must be evaluated with each failure to determine the optimum course of action. These repair-versus-discard decisions become increasingly difficult as the cost of repair approximates the cost of replacement, and they require consideration of additional, and frequently subjective, factors. Some of these factors are

- A. Availability of a suitable replacement. A shortage of an item may indicate repair is desirable when it would normally not be viewed as economically feasible or practical.
- B. Technological advances. The new product may incorporate increased sophistication, which would indicate replacement as the preferred choice, although repair may be practical.
- C. The stage of the product's useful life. The failed product may have been in service for an extended period of time. Expectations of more frequent failures in the future may lead to a replace decision when repair is cost effective.
- D. A desire, sometimes irrational, for a new product. The failure may provide an excuse to replace an otherwise repairable item.

The preceding list of factors clearly indicates that repair-versus-discard decisions are, to some degree, intuitive and based on individual analysis of each item that is potentially repairable.

Levels of Maintenance

Once the decision to repair a product is made, where the repair work should take place must be determined. This is derived by analyzing the various levels of maintenance capability and their availability at alternative locations.

Maintenance capability is subdivided into three levels of maintenance: organizational, intermediate, and depot. The three levels of maintenance are determined by the knowledge required for performance of the activity, the degree of sophistication inherent in the task, and the magnitude and type of support equipment required for maintenance

Organizational maintenance is the least sophisticated level of maintenance, and it requires little specialized knowledge and little, if any, maintenance support equipment. For example, consider the electrical system of a typical house. Replacement of a defective light bulb is a relatively simple task, requiring little specialized knowledge and no equipment. This task is well within the capability of the average homeowner.

Next, consider the replacement of a defective light switch. This task is more sophisticated and requires some knowledge of electrical wiring procedures and practices and some maintenance support equipment. This is an example of *intermediate maintenance*.

Continuing with this example, assume that the homeowner wishes to add a new circuit breaker with associated switches and wiring. The task is now more sophisticated and requires a high degree of specialized knowledge and a more complex array of maintenance support equipment. It is an example of *depot maintenance*.

It is important to note that the three levels of maintenance refer to an activity rather than to a specific location or maintenance facility. Any level of maintenance can be performed at any location (facility), or the three levels can be performed at different locations.

This division of maintenance into three levels is partially subjective. As knowledge becomes more specialized, more complex tasks tend to shift toward less sophisticated maintenance levels.

Repair-Level Decisions

Repair-level decisions are based on optimum repair-level analysis (ORLA) or level of repair analysis (LORA). Each technique is a method of establishing cost-effective repair policies while maintaining adequate support effectiveness.

LORA (or ORLA) begins with decisions concerning what should be repaired and initial conclusions regarding the optimum maintenance level (organizational, intermediate, or depot). The results of this analysis must then be balanced against additional criteria such as

A. The skills and knowledge of the intended user. Providing the requisite tools, test and support equipment, documentation, and inventory of spare and repair parts is of no value if personnel capabilities prohibit their use.
B. The quantity and type of maintenance support equipment that is to be provided. The level of repair at any location cannot exceed the capabilities provided by the maintenance support equipment which supports the activity.
C. The inventory of spare and repair parts. Properly trained personnel with the appropriate maintenance support materials can effect repair. This capability is meaningless, however, if the spare parts needed to restore service are not available.

A balanced and realistic integration of these factors is vital to effective and efficient product support. This can only be realized through the total involvement of logistic specialists throughout the product development phase. The product so designed is an ideal candidate for repair level analysis to determine the optimum degree and type of maintenance support.

Life Cycle Cost

Life cycle cost (LCC) is, quite simply, the cost of obtaining, operating, and sustaining a product, that is, the total cost of ownership. The LCC of a new product incorporates a pro rata share of design and development into the acquisition cost. Added to this are the costs associated with normal operation and all costs related to maintenance support. As an example, consider the following cost elements that make up the LCC of a typical automobile:

- A. Acquisition cost, which includes a share of research and development, advertising, sales commissions, warranty costs, dealership facility costs, and a host of other costs into the consumer purchase price.
- B. Daily operating costs including fuel, insurance, and regular, scheduled maintenance.
- C. Licensing costs for both the operator(s) and the vehicle.
- D. Replacement costs for items such as tires.
- E. Repair costs including loss of income when a malfunction impacts earning potential.
- F. Interest charged on the loan used to purchase the vehicle.
- G. The opportunity cost of other possible uses for the money used in purchasing the automobile.

This example clearly illustrates some of the elements of LCC that apply to the automobile. Figure 4.4 is another illustration of the incremental cost elements that make up the LCC. At first glance it may appear that this figure is applicable only to large military contracts and the acquisition of major weapons systems. Typical consumers don't pay for LCC incremental costs such as development training, technical publications, and spare parts, or do they? To explore the potential merit of this statement consider the consumer who purchases a personal computer.

- A. The manufacturer may incorporate improvements or enhancements into the product. The development costs of these improvements or enhancements are factored into the selling price and borne by the consumer.
- B. The computer includes, at a minimum, an instruction manual. The selling price includes the cost of this technical documentation.
- C. The consumer may desire a line conditioning system (support equipment); extra floppy disks or printer ribbons (operating spares); training on general computer usage or specialized programs. All of these are either included in the product price or purchased as "extras."

The concept of LCC is applicable to all products, although all elements of LCC may not apply to every item. For example, the logistic element of training may not

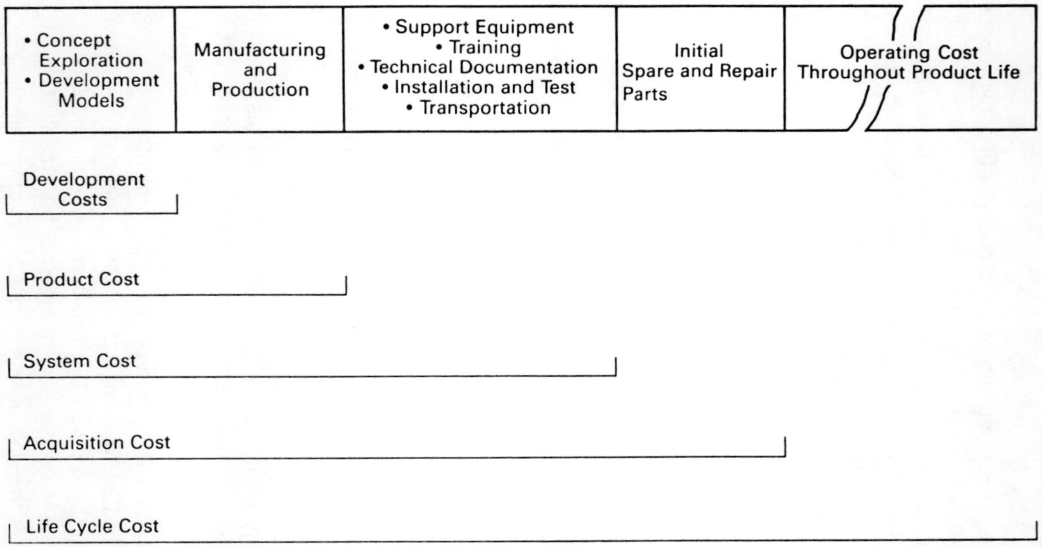

| Concept Exploration • Development Models | Manufacturing and Production | • Support Equipment
• Training
• Technical Documentation
• Installation and Test
• Transportation | Initial Spare and Repair Parts | Operating Cost Throughout Product Life |

|─ Development Costs

|─────── Product Cost

|──────────────── System Cost

|─────────────────────── Acquisition Cost

|── Life Cycle Cost

Fig. 4.4. Incremental Life Cycle Cost Elements

apply to the use of a flashlight; yet, a small inventory of spare batteries (another logistic element) may represent an intelligent choice.

LCC cost elements frequently far exceed the initial purchase price. Intelligent utilization of reliability engineering, maintenance engineering, and repair level analysis holds the potential of significant cost reductions within the logistics support arena.

Even a relatively small improvement in product design may lead to dramatic cost savings. For example, an average increase in fuel economy from 25 miles per gallon to 26 miles per gallon would reduce fuel consumption from 600 gallons to approximately 577 gallons for the automobile driven 15,000 miles per year. At a fuel price of $1.30 per gallon, this reduction in consumption saves $29.90 per year per vehicle. Assuming a reasonable sales figure of 8.5 million vehicles, the total yearly savings to the consuming public is a staggering $254,150,000!

Logistics is primarily concerned with support costs where equally dramatic savings are possible through the efforts of the reliability engineer, the maintainability engineer, and repair level analysts. The reliability engineer who decreases the failure rate (increases the MTBF) decreases support costs through reducing the need for service. The maintainability engineer who develops a product that is easier to repair reduces support costs since less repair time and equipment are required. Repair-level analysis enhances support productivity by providing the right part at the right place at the right time and in the right quantity, thereby optimizing support costs. LCC, in turn, permits measurement of long-term support costs, thereby placing what may appear to be high logistic development costs into the proper perspective.

SUMMARY

Logistics is properly designated as a support function. It provides support to the firm during the transformation process (operations logistics) and to the products of the firm following the point of ownership transfer (systems logistics). Support, whether to the firm or to the product, is represented by the various elements of logistics. The ILS manager attempts to integrate these logistic elements into a coherent and harmonious package of logistic resources that support and complement management objectives.

The ILS manager is somewhat limited because logistic support capability, or logistic supportability, is largely a function of product design. The prerequisites to logistics—reliability, maintainability, repair level analysis, and life cycle costs—must be considered throughout the product design and development process if logistic support is to be optimized.

Reliability indicates the probability that a given product will continue to function as intended, over a stated period of time (cycles, miles, events, and so forth), and under defined conditions. The most common units of measure are the MTBF and the failure rate.

Products do fail. For this reason maintainability, or the relative ease of repair, becomes an important logistic criterion. The product that is easy to repair can be returned to service quickly, thus decreasing out-of-service expense.

The reliable and maintainable product must next be analyzed to determine the appropriate level and optimum location for maintenance. The life cycle costs (LCC) may then be used to verify that the selected logistic support package is, in fact, optimum.

QUESTIONS FOR REVIEW

1. The text provided one example of logistics operating within the constraints of an existing product. Identify at least two other examples.
2. Explain why reliability is referred to as a prerequisite to logistics.
3. What is the difference between reliability and maintainability?
4. What are the advantages of repair-level analysis?
5. What is the relationship between MTBF and the failure rate?
6. A product experiences 8.5 failures in 11,000 hours of operation. What is the failure rate and MTBF?
7. What is mean downtime?
8. Give an example of net maintenance time.
9. Describe the three levels of maintenance.
10. Define the life cycle cost concept.

5

LOGISTIC CONNECTIVITY

No firm can exist as a self-sufficient entity. Each firm creates products via the transformation process through the utilization of internal resources. These internal resources in turn require access to resources external to the firm if the transformation of incoming materials into finished products is to be sustained. Transformation, without a continuous input of materials for production, will quickly become an impossibility. Similarly, the products of production must be moved from the firm at a rate approximating the length of the transformation process, or the firm will cease to exist as a viable enterprise. Firms receive material from and they supply products to the environment in which they operate. They must exist in consonance with this environment, which is inexorably linked to their survivability and growth.

This relationship may be represented by considering the firm as one leg of an interdependency triangle (see Figure 5.1) in which the supplier, the firm, and the marketplace interact to create a dynamic and interdependent system. All elements of this system must work together in harmony and coherence if the firm is to grow and prosper. Suppliers and the marketplace are components of the environment, and the effectiveness of their support determines the success or failure of each firm. The firm looks to this environment for the resources of production and the consumption of finished goods.

Materials for production must be transferred from the supplier(s) to the firm. Finished goods must be transferred from the firm to the marketplace. This creates an interactive unit made up of the supplier, the firm, and the marketplace. These three components must be connected together if the firm is to operate as part of a

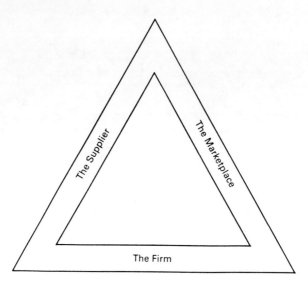

Fig. 5.1. The Interdependency Triangle

functioning system. Transportation provides this logistic connectivity by linking the firm to both the supplier and the market.

Transportation of material between suppliers and the firm and of products between the firm and its customers represents a major industrial activity of enormous proportions. The significance of this activity is illustrated in Table 5.1.

Transportation also plays a key role in supporting the products of production. Transportation resources are required, for example, in the movement of personnel and materials from the firm (or from an intermediary specialist) to the various areas where maintenance and repair are performed. Alternatively, products requiring repair may also need to be transported from the consumer to a repair facility; again requiring transportation. Transportation is also involved in establishing the initial inventory of spare and repair parts and their replacement.Transportation is a critical resource of the firm and frequently represents the most expensive logistic element. Requirements for transportation do, however, widely vary from industry to industry. For example, hundreds of railcars are required to transport a week's production of an automobile assembly plant, whereas United Parcel Service or the U.S. Postal Service suffice for the small mail-order business.

Transportation Modes

Transportation can be categorized into five basic modes: road, rail, water, air, and pipeline. Each mode incorporates specific advantages and disadvantages that determine its usefulness within any given industry. There is no "best" mode for a

Table 5.1. VOLUME OF DOMESTIC INTERCITY FREIGHT BY TYPE OF TRANSPORT: 1960 TO 1985

Type of Transport	1960	1965	1970	1973	1974	1975	1976	1977	1978	1979	1980	1981	1982	1983	1984	1985
	\multicolumn{16}{c}{Billions of Ton-Miles[a]}
Railroads[b]	579	709	771	858	852	759	800	834	868	927	932	924	810	841	935	893
Motor Vehicles	285	359	412	505	495	454	510	555	599	608	555	527	520	575	605	605
Inland Waterways[c]	220	262	319	358	355	342	373	368	409	425	407	410	351	359	382	353
Oil Pipelines	229	306	431	507	506	507	515	546	586	608	588	564	566	556	568	559
Domestic Airways[d]	0.9	1.9	3.3	4.0	3.9	3.7	3.9	4.2	4.8	4.6	4.8	5.1	5.14	5.87	6.58	6.4
Total	1,314	1,638	1,936	2,232	2,212	2,066	2,202	2,307	2,467	2,573	2,487	2,438	2,262	2,333	2,497	2,416
	\multicolumn{16}{c}{Percent Distribution}
Railroads[b]	44.06	43.28	39.82	38.44	38.52	36.74	36.33	36.15	35.18	36.03	37.5	38.1	36.0	36.0	37.5	36.9
Motor Vehicles	21.69	21.92	21.28	22.63	22.38	21.97	23.16	24.06	24.28	23.64	22.32	21.7	23.1	24.6	24.2	25.04
Inland Waterways[c]	16.74	16.00	16.48	16.04	16.05	16.55	16.94	15.95	16.58	16.52	16.37	16.82	15.6	15.4	15.3	14.61
Oil Pipelines	17.43	18.68	22.26	22.72	22.88	24.54	23.39	23.67	23.75	23.64	23.64	23.2	25.1	23.8	22.7	23.13
Domestic Airways[d]	0.07	0.12	0.17	0.18	0.18	0.19	0.18	0.18	0.19	0.18	0.19	0.21	0.23	0.25	0.26	0.26
Total	100.0	100.0	100.0	100.0	100.0	100.0	100.0	100.0	100.0	100.0	100.0	100.0	100.0	100.0	100.0	100.0

[a] The ton-mile is the basic unit of measure in the movement of materials and products (freight). A ton-mile is the movement of 1 ton (2,000 pounds) of freight over a distance of 1 mile.
[b] Includes electric railways and rail commuter service.
[c] Includes Great Lakes.
[d] Revenue service only for certified route and charter carriers, with small Section 418 all-cargo carriers included from 1978. Also includes express, mail and excess baggage.
[e] 1985 Figures are estimates

Source: Courtesy of Transportation Policy Associates, Washington, D.C., *Transportation in America*, March 1985. Used with permission.

given firm. The preferred choice is a function of many factors such as (1) the type of industry, (2) the location or locations of the firm, (3) the location and distribution of suppliers, (4) the marketing area, and (5) the availability of various transportation modes. A given firm may use one, all, or any combination of the five basic modes.

Care, however, must be exercised when selecting a specific mode. A choice based upon the lowest transportation charge may, for example, adversely affect total logistic costs. Such decisions must be coordinated throughout the logistics function to avoid choosing a slow mode of transportation because of its low cost. The lack of speed might require the construction of new warehouses, the establishment of huge inventories, or other expensive logistic alternatives. There are numerous options available, and the ILS manager must evaluate potential choices.

Road

Road (highway) transportation began to rise in prominence shortly after World War II. The growing popularity of this transportation mode was considerably aided by its virtually unlimited scheduling flexibility and its potential for door-to-door delivery. Road transportation received an even greater boost in 1956 when the Federal Aid Highway Act inaugurated the interstate highway system.

The increasing availability of more and better roadways paralleled a phenomenal increase in intercity freight tonnage transported by over-the-road motor carriers. By 1985, the estimated annual expenditures for this activity exceeded $207 billion! (See Table 5.2.) This is greater than the combined total for all other transportation modes.

Table 5.1 reveals that motor carriers accounted for over 600 billion ton-miles in 1983, or approximately 25 percent of all intercity freight. Future prospects for motor carriers appear good, and they are expected to maintain this approximate percentage.

Motor carriers, however, are not without problems. The fuel crisis during the early and late 1970s resulted in greatly increased transportation costs, leading to intense investigations into means for increasing the economy and efficiency of motor carriers. The rise in fuel costs, however, had an approximately equal impact on the economy as a whole, which quickly adjusted to increased costs of doing business. More significant items increasing costs in the industry are wages, equipment replacement, maintenance, and dock and platform charges. Although accelerating wage rates impact all modes of transportation, the effect is most severe in the labor-intensive motor carrier industry. This trend has been partially offset through improved scheduling, advanced billing practices, tandem units (two trailers pulled by a singe prime mover), coordinated transportation systems, and the emergence (though slow) of mechanized terminals.

Table 5.2. FREIGHT TRANSPORTATION OUTLAYS BY TYPE OF TRANSPORT: 1960 TO 1984

Type of Transport	1960	1965	1970	1974	1975	1976	1977	1978	1979	1980	1981	1982	1983	1984
Highway	$31.5[a]	$43.8	$62.5	$ 84.8	$ 84.8	$ 98.0	$111.4	$128.8	$142.7	$155.4	$170.9	$175.2	$189.5	$207.1
Truck: Intercity	18.0	23.6	33.6	48.8	47.4	56.2	67.4	79.6	90.2	94.6	101.7	102.7	110.6	118.5
ICC Authorize[b]	7.2	10.1	14.6	22.7	22.0	26.0	31.0	36.5	41.2	43.0	47.9	44.1	46.5	52.1
Non-ICC Authorized[b]	10.2	13.6	19.0	26.1	25.4	30.2	36.4	43.1	49.0	51.6	49.9	55.8	64.6	66.4
Truck: Local	13.5	20.1	28.8	35.9	37.3	41.6	43.9	49.1	52.3	60.5	72.8	75.0	78.1	88.4
Rail	9.0	9.9	11.9	17.0	16.5	19.2	20.5	21.9	24.8	27.7	30.5	27.1	27.3	30.1
Water	3.3	3.8	5.1	8.0	7.9	8.9	10.3	12.4	13.4	15.5	16.6	15.6	16.2	17.9
Oil Pipeline	0.9	1.1	1.4	1.9	2.2	2.5	3.2	5.5	6.2	7.1	7.4	7.9	8.3	8.6
Air	—	—	—	—	—	—	—	—	—	4.0	4.3	4.4	4.9	5.7
Freight, Total[c]	46.8	61.1	83.8	115.7	115.5	133.2	150.4	174.0	192.9	212.7	233.1	235.4	249.5	272.9

[a] Figure are in billions of dollars.
[b] ICC is the Interstate Commerce Commission.
[c] Includes items not shown separately such as outlays for mail and express.

Source: Courtesy of Transportation Policy Associates, Washington, D.C., *Transportation in America*, July 1985. Used with permission.

Rail

Railroads have a rich heritage and historically have accounted for the largest share of intercity freight. A comprehensive rail network interconnecting virtually every city permitted the railroads to establish a quasimonopolistic position in the industry. Railroads dominated intercity freight movement through World War II. Increasingly intense competition following World War II, led by the growing motor carrier industry, resulted in significant declines in the railroads' share of intercity ton-miles. This share dropped from a high of 54 percent to approximately 36 percent in 1985 (see Table 5.1). Railroads do, however, offer a significant advantage: They can transport an enormous tonnage, and their share has stabilized at approximately 36 percent.

As a result of the increased competition, the railroads mounted a determined effort to discontinue low-volume shipments and rail service to small cities. This led to a steady decline in the number of rail miles. The number of rail miles is currently less than 200,000, ranking it fourth in the five available modes. Only inland water transportation accounts for a lower number of transportation miles.

Water

Water is the oldest form of transportation, and the availability of this medium led to the establishment of major cities near navigable bodies of water. The primary advantage of water transport is its ability to carry vast tonnage at relatively low freight rates. This advantage will continue to place water transport in demand whenever speed of delivery is not a consideration.

The primary disadvantage of water is its inflexibility. Roads, railways, airline terminals, and pipelines may be constructed as desired within rather broad limits. Waterways, on the other hand, either exist or they do not. A firm wishing to utilize water as a transport medium must be located adjacent to the waterway or encounter the additional costs of supplemental transportation modes such as motor carriers or railroads. This severely limits the available locations, as there are only approximately 26,000 miles of navigable inland waterways in the United States. Thus, water with the least number of miles is the smallest of the five transportation modes.

Air

Air is the newest yet least utilized medium for the transport of freight because cargo limitations, lift capability, and cost effectively reduce its availability. Another disadvantage is the need for an additional transportation mode between the firm and the air terminal. The overriding advantage of air is its speed of delivery. Coast-to-coast shipments between major cities are measured in hours instead of days. Speed offers the potential of profitable trade-offs with other logistic functions. For example, delivery within hours could lead to reduced warehousing requirements. The resultant savings could more than offset the increased cost of air freight.

Air transport, however, is still very much a potential rather than a reality. Air freight is largely relegated to nonroutine and emergency shipments or those that have high dollar values or are highly perishable. Total intercity freight tonnage moved by air is less than 1 percent (see Table 5.1). This small freight volume has a limiting effect on the amount of money available for the research and development of advanced techniques in moving air freight. The cycle is thus self-perpetuating.

Nevertheless, the future of air freight within the logistics arena appears promising. Innovations such as the move toward hub-and-spoke systems during the early 1980s have revolutionized the air cargo industry. In a hub-and-spoke system, a central terminal (the hub) is fed by trucks (the spokes). This necessary "marriage" between the truck and the airplane initiated a trend toward a total transportation company where a single call serves the needs of a shipper.

Although air transport is expensive and requires the use of supplemental transportation resources, the speed of service may well reduce total logistic costs. It is in this area that movement by air holds the greatest promise.

Pipeline

Pipelines first appeared during the latter half of the nineteenth century and, following decades of relatively slow growth, accounted for just under 10 percent of intercity freight tonnage by the 1940s. The use of pipelines dramatically increased following World War II, and today their use commands approximately 23 percent of the market (see Table 5.1). A significant portion of this growth stems directly from government subsidies instituted following the energy crisis of the 1970s.

There are approximately 250,000 miles of pipelines in use in the United State. The transport of petroleum accounts for the most use. Construction costs, right-of-way acquisition costs, and the cost of control stations give pipelines the highest fixed cost of all transportation modes. This cost, however, is partially offset by lower operating costs resulting from largely unattended, nonstop operation.

The major disadvantage of pipelines is the severe restriction of the type of commodity that can be transported. Recent research efforts have investigated the possibility of moving additional products in the form of slurry or in hydraulic suspension. However, petroleum will remain the primary pipeline commodity for the immediate future.

Legal Groupings of Carriers

The transportation alternatives listed above are grouped on the basis of the specific mode. Another frequently used grouping is related to the *legal definition* of the operating rights that govern each carrier. The four basic legal types are the common carrier, the contract carrier, the private carrier and the exempt carrier. Each type may exist within any of the five basic transportation modes.

Common Carriers

The *common carrier* is the most frequently employed legal category for transportation resources. A common carrier is a firm that transports for revenue at any time and to any place within its operating jurisdiction. They are required to publish all rates charged for this service, and the rates charged must be the same for similar services. Common carriers are authorized to offer transport for hire upon receiving a certificate for public convenience and need.

Prior to 1980, restrictions virtually eliminated any variation or flexibility in carrier operations. A shipper purchased approximately the same level of service from each carrier and was charged an almost identical price for this service. Deregulation, one result of the Motor Carrier Act of 1980, removed a lot of these restrictions and led to a larger range of service and price options.

Contract Carriers

The *contract carrier* receives a permit that authorizes the transportation of specific items over specified routes. Transportation services provided by the contract carrier arise from contractual arrangements between two parties: the shipper and the carrier. The contract provides the shipper with a defined transportation service at a mutually agreeable price. Contract carriers, unlike the common carrier, are not required to charge the same rate for equal service.

Exempt Carriers

Exempt carriers do not fall under the umbrella of direct regulations with regard to pricing policies and operating rights. They do, of course, have to conform to the laws of the state in which they operate. The exempt carrier initially transported commodities such as agricultural products to processing centers. Currently exempt carriers support a much larger range of activity. Exemptions may also be granted for specified areas such as within a city and the contiguous commercial areas of activity.

Private Carriers

Private carriers originally consisted of transportation resources that are controlled by the firm through ownership or lease. This was changed in the early 1980s when a federal court cleared the way for private carriers to use owners-operators or other outside sources of vehicles and drivers. Private carriers are restricted in that the material being shipped must be owned by the firm and the transportation of that material must be incidental to the primary business of the firm. These carriers are enjoying a period of sustained growth due to the dual advantages of increased flexibility and greater economy of operations. These advantages cannot be duplicated by the other, more general purpose carriers.

Multimode Systems

Multimode systems incorporate a method of meeting transport objectives through a combination of two or more transportation modes. Theoretically, multimode systems could result from any combination of the five basic modes. In practice, however, specific combinations have evolved, and terms such as *piggyback, fishyback, airtruck,* and *trainship* are now part of the transportation vocabulary.

The best known and most common multimode system is the *piggyback* or, to use the more correct title, *trailer on flatcar* (TOFC). As implied by its name, TOFC refers to a motor carrier trailer that is placed aboard a railroad flatcar for a portion of the route.

One of oldest types of multimode transportation is *fishyback.* In fishyback, a motor carrier trailer or a shipping container is placed aboard a ship or barge for a portion of its journey. *Containers* are protective enclosures in which goods are stored during transit. The container facilitates handling and provides a measure of protection to its contents. Containers are also shipped via flatcar, in which case they are referred to as a *container on flatcar* (COFC).

The use of multimode transportation systems promises greater economies and increased efficiency. The spirit of deregulation introduced by the Motor Carrier Act of 1980 greatly enhanced multimode potential by permitting the so-called supermarket service. Under this concept, carriers can offer shippers a total service embracing all modes of transportation.

The benefits of multimode transport can be obtained through determining the optimum transportation method for each portion of the projected route. Logistics has the responsibility of evaluating the available options and selecting those best suited to the cost-effective and efficient use of transportation resources.

Regulating the Common Carrier

Common carriers represent the most versatile legal form of transportation because they are permitted to transport virtually any commodity to any location within a defined operating area. The common carrier must offer this transportation service without discrimination and, until 1980, had to charge the same rates for similar shipments. This freedom of commodity content and movement led to the common carrier being subject to complex and direct authorizations governing operating rights, transportation charges, and safety regulations.

All legal forms of transportation are subject to some degree of regulation that acts to scrutinize their activities in the public interest. This control was the province of individual states until passage of the Act to Regulate Commerce in 1887. This act marked the beginning of the federal government's regulatory power and created the Interstate Commerce Commission (ICC).

The federal government's increasing role as a regulatory power was subjected to additional definition during the early part of the twentieth century. The Elkins

Act (1903) reduced rebates and special concessions and increased the penalties for deviating from published rates. This was followed by the Hepburn Act in 1906, which expanded the just-and-reasonable review authorization of the Act to Regulate Commerce (1887) to include examinations of maximum rate levels. The Hepburn Act, however, was largely impotent until passage of the Mann-Elkins Act in 1910. This act allowed the ICC to issue rulings on the reasonableness of proposed rates and to suspend them when they were deemed discriminatory.

Modern rate regulation was completed in 1920 with the passage of the Transportation Act. This act expanded the authority of the ICC by permitting it to prescribe both minimum and maximum reasonableness of rates. The ICC was also directed to assume a more aggressive role regarding proposed rates. The Transportation Act of 1920 changed the name of the Act to Regulate Commerce to the Interstate Commerce Act and modified it to permit initiation, modification, and adjustment of rates as necessary and in the public interest.

The Emergency Transportation Act (1935) provided additional refinement by instructing the ICC to set standards with respect to reasonable rate levels. In that same year, the Motor Carrier Act of 1935 placed regulation of common carrier highway transportation under the jurisdiction of the ICC. This act, known as Part Two of the Interstate Commerce Act, defined the legal forms of common, contract, and exempt motor carriers. Several other transportation related acts were implemented to define and clarify issues arising from the basic acts of 1887 and 1920.

The Motor Carrier Act of 1980 represents a significant milestone in highway transportation. This act resulted in partial deregulation, thereby permitting carriers to set special tariffs. A firm could now negotiate reduced transportation charges for the materials and products of production.

It should be noted that the ICC does not determine or set rates for carriers falling under its jurisdiction. The function of the ICC is to review and either approve or disapprove rates that have been established by the carrier.

Transportation Rate Structures

Common carriers employ two major systems in deriving rate charges for shipments. In the first system, all products transported by common carriers are divided into a relatively small number of classes. Rates, referred to as *class rates*, are then quoted for the particular class instead of the individual items. This avoids the unmanageable alternative of separate rates for each of the thousands of items capable of being transported. In the second system, common carriers quote rates on specific individual items instead of using the rate classifications. These are referred to as *commodity rates*.

A third type, exceptions to the classification, is a modification of the class rate. This *exception rate* was established to permit special charges for shipments within a specific area or of a special commodity when either competition or volume provided justification for a rate adjustment. Exception rates are normally thought of as reductions from class rates, although they may be higher or lower. For example, low volume may justify an exception rate that is above the class rate.

Class Rates

Class rates group all products that are normally transported into uniform classifications. Classifications, based on product characteristics that influence the cost of handling or transport, afford significant reductions in the range of transportable products. The particular class that an item receives is its *rating*. The rating is not the charge incurred in transporting the product, although, in general, products having a higher rating are subject to higher transportation charges. The actual prices paid for transportation are called *freight rates* and are published in pricing sheets called *tariffs*.

Factors such as value, susceptibility to theft or damage, item density, and type of packaging exert an influence on item classifications. For example, a product that is highly desirable, valuable, and susceptible to theft during transport (such as the personal computer) may receive a higher classification than a similar product having less value or less desirability. A high susceptibility to damage during transport may also lead to a higher classification, since the carrier's cost varies directly with damage, which in turn affects the rate.

Products are also grouped as a function of the quantity being shipped. For example, less-than-carload lots (LCL) or less-than-truckload lots (LTL) have higher transportation charges than carload (CL) or truckload (TL) lots.

The first step in determining transportation charges is to identify the class rating. The class rating is then used with the tariff to determine the applicable charges per hundredweight (equal to 100 pounds) between any two points. The actual price is, however, normally subject to a specified minimum charge, which is imposed without regard to weight and represents the lowest price a shipper pays for transportation.

The actual price may include a surcharge or an arbitrary. A *surcharge* is an additional charge added to defray the cost of handling. It is frequently applied to small shipments and may be either a flat charge or a sliding scale based upon size. An *arbitrary* is a special assessment added to transportation charges. It is added to provide special compensation to the carrier for service to a specific destination or area.

Class rates, minimum charges, surcharges, and arbitrary charges make up a general rate structure. This rate structure is, in various combinations, applicable to the movement of goods between any two points within the contiguous United States.

Commodity Rates

A *commodity rate* is a special rate listed directly in a tariff. It takes precedence over any class rate that may be identified for that same commodity. Commodity rates were developed to accommodate the regular movement of large quantities of a product on the theory that this would encourage volume shipments. Commodity rates normally apply only to the products specified in the tariff and are issued on the basis of transportation between two successive points.

Special Rates

The rates described above are supplemented by a variety of *special rates,* many of which are of interest to logistics and the ILS manager. The first of these special rates is the *freight-all-kinds rate* (FAK). FAK permits the delivery of a variety of commodities to a limited number of destinations. Charges are based on an average for the total shipment instead of a separate charge for each classification. FAK is of particular importance to the warehouse that carries an assortment of goods for delivery to one or more consumer outlets.

Another type of special rate is the *local rate,* which is imposed on an item that moves under the tariff of a single carrier. When the transportation involves a single shipping document (bill of lading) and multiple carriers, it is referred to as a *joint rate.* The joint rate may offer significant savings over multiple local rates and thus should be considered by the logistician when evaluating transportation options.

Proportional rates, another classification of special rates, offer price incentives for using a tariff that applies to a portion of the transport route. They are normally applied to shipments (either origin or destination) that extend beyond local tariff boundaries. The proportional reduction over a simple extension of the local tariff may result in lower total transportation charges.

Another alternative, the *combination rate,* is a special rate obtained by using any combination of class, commodity, or exception rates. Combination rates may be employed when no published local or joint rates exist between two locations.

SUMMARY

Transportation provides logistic connectivity through the movement of materials into the firm and products from the firm. It is a critical resource of the firm and frequently represents the most expensive logistic element.

Transportation encompasses the five basic transport modes of road, rail, water, air, and pipeline. Transportation assets are further divided, by legal description, into common, contract, private, and exempt carriers. Each of the five basic modes may exist as any of the four legal carrier types. A given firm may use a single mode or any combination of modes to obtain logistic connectivity. It is the responsibility of logistics and the ILS manager to select the optimum mix of transportation resources for the firm.

Which resource is optimum, however, cannot be determined solely by an analysis of costs per ton-mile. Transportation needs must be evaluated within the context of the firm with the cost per ton-mile measured against potential impact to the firm and the total logistic cost.

The common carrier rate structure is the means of evaluating the logistic cost-benefit ratio of various transportation possibilities. These data may then be used, in conjunction with the transportation needs of the firm, to determine the modes that most closely approximate management goals. This analysis is facilitated by carriers that offer a total transportation approach and "supermarket" service.

QUESTIONS FOR REVIEW

1. What are the five basic transportation modes?
2. What are the legal characterizations of carriers?
3. Explain how any one of the transportation modes can function as any one of four carrier types.
4. What percentage of intercity freight is moved by motor carrier?
5. Define the concept of logistic connectivity.
6. Differentiate between a common carrier and a contract carrier.
7. What is a tariff?
8. What is an arbitrary?
9. What is the basis for class groupings?
10. Define commodity rate.

6

LOGISTIC FACILITIES

The successful accomplishment of logistics objectives requires the development of structures specifically designed to enhance the logistics function. These structures, or logistic facilities, encompass a variety of styles and serve numerous functions. Warehouse facilities are required for the storage of incoming raw materials used during the production process. As few products are made to order, additional warehouse space is required to store the finished product until it has been sold. Facilities are also required for the storage and control of inventory items used in supporting the equipment of production. Inventory and the attendant inventory storage facilities are also required to support the product following a transfer of ownership.

Additionally, facilities are required to support logistic activities such as training. For example, the manufacturer of complex military systems may be expected to train customer personnel in the operation and maintenance of equipment provided by the firm. Other examples include firms dealing in sophisticated consumer items such as the personal computer. Training consumers to use a product may provide a competitive advantage to the firm and increase customer loyalty. Still other facilities may be needed to support maintenance activities related to equipment within the firm and for support to products of the firm.

Logistic facilities may be owned by the firm, operated as an independent business (an intermediary specialist), or both. For example, the automobile dealership normally includes a maintenance and repair facility to support products sold by the firm and automobiles in general. This facility is in direct competition with independent garages that offer identical services.

An additional factor enters the picture when the spares and repair parts inventory is considered. The automobile dealership, to continue with the previous example, maintains such an inventory to facilitate the repair process. This same resource may also serve as a source of spare and repair parts for the competing intermediary specialist. This means that the automobile dealer must evaluate internal usage and external demand when developing a facility for spares storage.

It is apparent that logistic facilities serve a variety of diverse functions. There are, however, two overriding issues common to all facilities and of primary interest to logistics: (1) What is the size of the desired facility? (2) Where should it be located? The ILS manager must play a major role in deciding these issues.

A thorough analysis of how to determine the size and location of each type of facility is neither possible nor desirable within the scope of this book. Instead, this chapter examines those factors as they relate to maintenance and warehouse facilities. Relatively simple extrapolations from this information may then be used to obtain comparable data for other facility types.

The Maintenance Facility

An effective and efficient maintenance resource is a vital member of the logistics community. This resource, whether included as a functioning part of the firm or a service provided by an intermediary specialist, is a major contributor to successful operations. The firm producing a product requires maintenance to assure smooth operation of production equipment and for efficient and effective repair when malfunctions occur. The firm that sells products to the consumer requires maintenance to support its product.

A well-organized and -managed maintenance capability is a decided asset to the firm. It must, however, properly serve the level of maintenance defined by management. A capability that exceeds or fails to meet the objectives of the firm represents wasted resources and a potential subtraction from the profit objective. Conversely, a capability that meets the firm's objectives will likely enhance the profit potential.

Any maintenance capability that is capable of meeting intended objectives is, to a significant degree, dependent upon selection of an optimum physical structure. It must be of sufficient size to assure efficient and effective utilization of space, yet incorporate enough reserve capability to accommodate periodic increases in the expected workload. The maintenance facility must be sized to support the level of maintenance desired by the firm.

Sizing the Maintenance Facility

The single factor exerting the most influence upon facility size is the magnitude of expected workloads. Workload requirements provide a quantitative measure that can be translated directly into personnel and work space needs. In addition, the maintenance facility normally provides space for maintenance-related activities. These related factors contributing to the size of the overall facility include

A. Space for storage and administration of spare and repair parts and associated materials
B. Space for storage of any required tools and test and support equipment
C. Space for storage of support items such as equipment drawings, technical publications, and consumables
D. Office space that may be required in meeting personnel needs
E. Alternative functions such as training that may be served by the facility

The predominant factor—expected workload—is a function of the levels of maintenance being supported and the number of repair actions the facility is expected to perform. The levels of maintenance to be supported represents a management decision, whereas expected repair actions are derived through an analysis of the product(s) being supported.

Expected Repair Actions

A maintenance facility exists to meet the needs for maintenance at any combination of the three maintenance levels—organizational, intermediate, and depot. The facility that supports all three levels performs more maintenance actions than the facility that supports only one level. The expected repair activity seen by the facility then becomes a function of the level or levels of maintenance being provided and the total equipment or product population that is being supported.

Maintenance activity estimates are derived through an analysis of reliability data. For example, assume that a maintenance facility has been established to support product Y having a mean time between failure (MTBF) of 1,500 hours. This value means that the product is expected, on the average, to experience one failure during each 1,500-hour interval of operation. It does not, however, provide any indication how often the item will fail in a given unit of time. This requires the assumption of an operational scenario.

Assuming an operating scenario of 6 hours per day, 7 days per week; product Y will undergo 42 hours of operation each week or 2,184 operating hours per year. Dividing this by the MTBF of 1,500 hours reveals that product Y may be expected to fail 1.456 times per year. Thus, if only one unit of the product is being supported, the maintenance facility would expect to have approximately 1.5 repair actions per year. Eq. (6.1) shows this relationship.

$$\frac{2{,}184 \text{ operating hours per year}}{1{,}500 \text{ MTBF}} = 1.456 \text{ failures per year} \qquad (6.1)$$

Repair actions are also a function of the total population being supported. The assumption above must therefore be expanded to include an assumed quantity of the product that is to be supported by the facility, for example, 2,250 units. Each product unit is assumed to have an identical MTBF and operating scenario; therefore, the total number of expected product Y repair actions in one year equals the repair actions for one unit (1.456) times the total population (2,250), or 3,276.

A second product supported by this maintenance facility has an assumed MTBF of 2,350 hours, an operating scenario of 8,736 hours per yer, and a population of 3,690 units. Substituting these values in Eq. (6.1), gives 3.72 failures expected per year per product unit.

$$\frac{8{,}736 \text{ operating hours per year}}{2{,}350 \text{ MTBF}} = 3.72 \text{ failures per year} \qquad (6.2)$$

The corresponding number of expected failures per year for the total product population is 3.72 × 3,690, or 13,727 repair actions.

$$3.72 \times 3{,}690 \text{ (units)} = 13{,}727 \text{ repair actions} \qquad (6.3)$$

The total repair actions expected in one year is the sum of the repair actions expected for each product.

$$3{,}276 + 13{,}727 = 17{,}003 \text{ repair actions} \qquad (6.4)$$

Similar analyses for each product result in an aggregate total of all repair actions that the maintenance facility will expect to handle each year.

The figure 17,003 represents the number of times repair action (corrective maintenance) is expected to be required as a function of the total population being supported. It does not include periodic preventive maintenance, regular servicing procedures, or false malfunction indications (through operator error, for instance).

An alternate approach to obtaining the number of times corrective action is expected considers the total number of all maintenance actions. Regularly scheduled procedures and false malfunction indications are then extracted to determine the quantity of expected maintenance actions requiring product repair (corrective maintenance). To illustrate this alternative approach, consider product Y from the previous example. This product is expected to fail 1.456 times per year. If this product also required regular service following each 160 hours of operation, a total of 13.65 service procedures are required per year (2,184 operating hours per year divided by 160 operating hours between service intervals). These procedures represent maintenance actions but normally do not require repair actions (corrective maintenance). The total number of all actions is the sum of 13.65 (scheduled service operations) and 1.456 (expected repair actions), a total of 15.106 maintenance actions. This total is derived from the product, its operational scenario, and the number and frequency of scheduled service routines. The result is the mean time between maintenance actions (MTBMA). The MTBMA for this product is expressed in Eq. (6.5).

$$\text{MTBMA} = \frac{2{,}184 \text{ (operating hours)}}{15.106 \text{ (all maintenance actions)}} = 144.58 \text{ hours} \quad (6.5)$$

The MTBA is now modified by a reliability-maintainability derived maintenance factor to extract preventive maintenance actions, false malfunction indications, and so forth, resulting in the mean time between corrective maintenance (MTBCM).[1] MTBCM is roughly equivalent to the MTBF.

It is important to note that MTBF is the statistical average of the time between failures for a large population of the product being considered. A given unit within this population may fail a few moments after being placed in operation or operate successfully for a period of time greatly exceeding the MTBF. Maintenance predictions for a small population have very limited validity.

Workload Determinants

The analysis above describes the total number of expected repair actions. It does not, however, provide information concerning the expected workload, which is the product of the number of repair actions and the average time required to complete each action. As in calculations involving the MTBF, repair times per maintenance activity are an average of many repair actions. Unit times have little validity, as individual repair actions may require more or less time than a statistical average.

Facility workload is a function of the number of repair actions, which is, in turn, a function of the levels of maintenance being supported. The facility that supports two or more levels of maintenance sees more activity than the facility supporting only one level. MTBMA includes all maintenance actions at each of the three levels—organizational, intermediate, and depot. Assuming the figure of 17,003 repair actions identified in equation 6.7 does not include organizational maintenance activities, this total represents intermediate and depot maintenance actions. Of this figure, approximately 60 percent will be successfully repaired at the intermediate level with the remainder (40 percent) requiring depot-level repair. Therefore, in the text example,

$$17{,}003 \text{ repair actions} \times 0.60 = 10{,}202$$

at the intermediate level and

$$17{,}003 \text{ repair actions} \times 0.40 = 6{,}801$$

---

[1] The maintenance factor extracts those activities that do not require repair or corrective maintenance. It is a combination of all preventive (scheduled) maintenance routines, their frequency of performance, the time required for performance and false malfunction indications such as operator error. For the example illustrated in equation 6.5, the maintenance factor is approximately 10.

at the depot level. The repair facility providing both intermediate and depot levels of service would, of course, expect to see the total number of repair actions (17,003).

Each item received at the repair facility must undergo a sequence of events during the repair process. These events may be summarized as follows:

A. Receipt of the item at the facility, including incoming inspection, assignment to a repair station, and the associated paper work
B. Initial checkout and test to verify the status of the incoming item
C. Repair of the item
D. Inspection, checkout, and test of the item following repair
E. Packing for shipment and associated out-processing

Events A, B, D, and E are applicable to each item received at the facility and, on the average, require a constant expenditure of time. Although the actual time involved varies with the administrative procedures at the facility and numerous other factors, it is not unrealistic to estimate four hours per item for events A and B and six hours per item for events D and E.

Event C, repair of the item, is highly variable because the status of the malfunctioning item is an unknown factor. The items can, however, be grouped into several categories of probable repair actions. Additionally, estimates of the number of received items that fall into each category are available. The categories, and the percentage of items in each category, are as follows:

A. Received items that are beyond repair, which only require a decision concerning their proper disposition (8 percent)
B. Received items that require little or no repair (40 percent)
C. Received items that require piece part replacement, repair, and alignment or adjustment (38 percent)
D. Received items that require extensive repair or rebuild actions (14 percent)

The number of repairable items per category that may be expected at the repair facility can be determined with these data. Assuming a facility supporting depot-level maintenance and the 6,801 repairable items from the previous example, the numbers per category are as follows:

Category A: 6,801 × 0.08 = 544
Category B: 6,801 × 0.40 = 2721
Category C: 6,801 × 0.38 = 2584
Category D: 6,801 × 0.14 = 952

Maintainability engineering is then consulted to derive an average repair time for each product category. Repair times vary with the type of product being supported; therefore we continue this example using times representative of a depot facility repairing electronic systems.

Category A items are beyond repair. These items are detected during incoming inspection, and therefore consume no repair time. Total facility time consists of the 4 hours per item allocated to incoming repair times.

Category B items require little or no repair time. An overall average of 1.5 hours of repair time per item is typical for this category. Total time per item then becomes 4 hours in-processing, plus 1.5 hours repair, plus 6 hours out-processing, or 11.5 hours.

Category C items require piece part replacement of defective components and extensive alignment and adjustment. This procedure requires an average expenditure of approximately 3.5 hours per unit, 4 hours in-processing and 6 hours out-processing totals 13.5 hours per item.

Category D items require extensive rebuilding or overhauling, an average of approximately 9 hours per unit. Total time is 19 hours per item when in-processing and out-processing times are added.

This information is now used with the number of repairable items in each category to determine the expected repair hour requirements for the facility.

Category A: 544 items at 4 hours per item   = 2,176 hours
Category B: 2,721 items at 11.5 hours per item   = 31,292 hours
Category C: 2,584 items at 13.5 hours per item   = 34,884 hours
Category D: 952 items at 19 hours per item   = 18,088 hours

Total   86,440 hours

This figure (86,440 hours) represents the total estimated time required to repair defective products received by the depot maintenance facility. The next step is to estimate the number of repair personnel required.

Personnel Requirements

Personnel requirements are a function of the tasks that must be performed and the time required for their performance. The example has identified a variety of tasks requiring an estimated annual time expenditure of 86,440 hours. To determine the number of personnel, the work hours available for task performance must be considered.

The typical work year, consisting of 52 weeks at 40 hours per week, totals 2,080 hours. However, the average employee receives 2 weeks vacation per year; thus the normal work year is usually considered to be 50 weeks or 2,000 hours. This total of 2,000 hours is further reduced by holidays, unscheduled time away from the job, and sick leaves. These factors, and other task assignments that must be performed, reduce the actual work year that can be planned against any defined task to about 1,600 hours.

Thus, each full-time employee devotes approximately 1,600 hours per year to the job (excluding overtime). The required task time (86,440 hours) divided by the hours available per employee (1,600) yields the total number of employees needed to accomplish the estimated workload, as expressed in Eq. (6.6).

$$\frac{86,440 \text{ required task hours}}{1,600 \text{ hours per employee}} = 54 \text{ employees} \qquad (6.6)$$

This number represents the staff requirements needed for in-processing, repair, and out-processing only. It should be apparent that additional personnel are required for management and supervision, clerical and administrative support, inventory control, and so forth.

Personnel requirements, as derived from the expected workload, have a value beyond the determination size for the new maintenance facility. These data can also be used to evaluate the adequacy of current staffing or to determine if overtime (as opposed to new hires) can meet increased workloads.

Facility size is also a function of the type of products being supported, the magnitude and quantity of tools and test equipment required, and the levels of maintenance being supported. Figure 6.1 illustrates a typical depot-level repair facility specializing in electronic communications systems and equipment. The figure incorporates maintenance, administrative, and storage areas into an efficient and repair-dedicated facility. A comparable floor plan for repair of vehicles, personal computers, or other types of products or maintenance levels would, of course, vary somewhat from this diagram. Space requirements are similar, however, since the primary determinant is people. The employee must have room to perform the task in question.

The Warehouse Facility

The systems concept emphasizes the flow of material into the firm, semifinished goods through the firm, and finished goods from the firm. A macroviewpoint extends this concept to encompass suppliers to the firm and customers of the firm and is illustrated in Figure 6.2. The flow of material through this macrosystem is characterized by interruptions. The supplier and the firm obtain material in excess of immediate need to maintain an uninterrupted transformation process. This excess material must be stored at the input of the supplier (and the firm) until needed. The supplier and the firm produce in excess of demand, and the excess must be stored until needed. These repeated interruptions of flow through the macrosystem necessitate facilities for storage—the warehouse.

Storage facilities do not necessarily equate to a building specifically designed to accommodate the materials for production or the products of production. Consider, for example, an extraction industry such as steel. Iron ore (the material of production) is stored in a suitable open area until needed. Consider also the automobile industry, where the completed product is stored in a large parking lot pending shipment to a new car dealership. Regardless of the nature of the facility, the function of warehousing is to create time and place utility, thereby improving service and lowering the cost.

During the early part of the twentieth century, warehouses were considered as

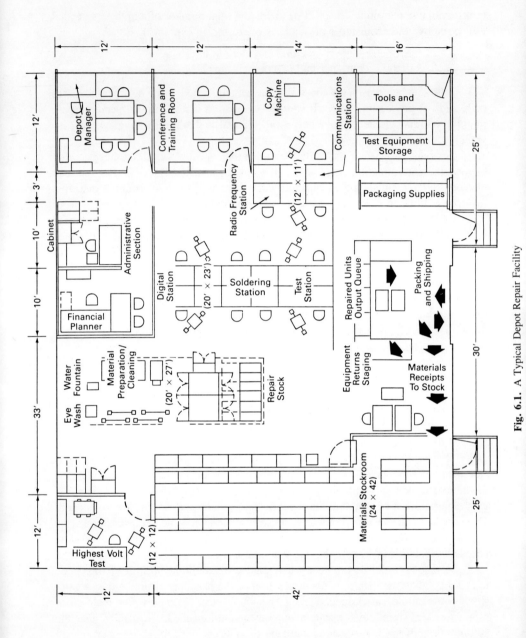

Fig. 6.1. A Typical Depot Repair Facility

Fig. 6.2. A Macroview of the System

providing only time utility—the matching of a relatively stable supply with an erratic demand. Finished goods or material were stored to assure availability when needed. No consideration was accorded the increased economies that were available through the use of alternate storage locations (place utility). Under this concept, materials and products were stored until the gradual build-up of demand justified distribution. Storage and the warehousing facilities necessitated by the requirement for storage were largely considered as necessary evils with little or no thought accorded to their potential value within the logistics function.

Low labor rates and an available work force contributed to this lack of interest by providing an unending supply of "arms and legs" for the repetitive movement and transfer of items stored within the warehouse. There was little incentive for seeking improvements in efficiency, space utilization, work methods, or material handling.

The move toward a consumer-oriented economy following World War II and the increased capabilities within the transportation arena focused managerial attention on increased operating efficiency. Place utility was added to the time utility of earlier warehouse operations.

The Addition of Place Utility

The emphasis on increasing warehouse efficiency led to a review of the function served by warehousing. Time utility is certainly a valid criterion, because products must be stored to accommodate seasonal variations and to bridge the gap between marketing and production. But was there another function to be served by the warehouse? The answer became quite clear as transportation capabilities increased, creating a growing awareness of the implications inherent in product distribution. The firm delivered small quantities of its product (relative to production) to either the individual or to retail dealers for subsequent resale (Figure 6.3). This operation required many trips, with each trip handling a small quantity of the product.

The new awareness of place utility led to the establishment of warehouses near the point of consumption (Figure 6.4). The firm could now make less frequent and larger shipments to the remotely located warehouse. Distribution of smaller quantities was made by more cost-effective, local transportation.

Retail outlets typically require a small quantity of a large variety of products. This characteristic led to the creation of the public warehouse (intermediary specialist), which stores an assortment of goods comprised of the products of many firms (Figure 6.5). The creation of these intermediary specialists increased the efficiency of transportation, since the local carrier could now deliver full loads, comprised of a variety of products, to several retail outlets.

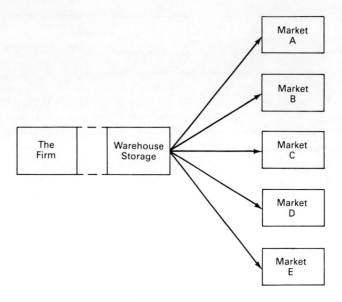

Fig. 6.3. Time Utility

Warehouse Functions

The warehouse receives large quantities of a product from the firm and provides smaller quantities of that product to other firms. Performance of this task necessitates the warehouse functions of movement and storage. Movement is divided into the four handling activities: reception, transfer, selection, and shipment.

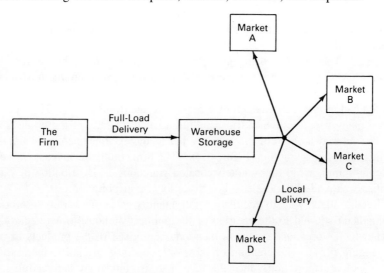

Fig. 6.4. Time and Place Utility

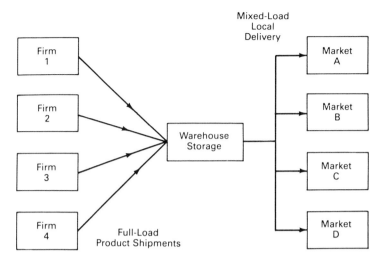

Fig. 6.5. An Assortment of Goods

A. Reception. Products and material are normally received at the warehouse in large quantities. The first movement activity is to unload incoming items. This is primarily a manual activity, although mechanization in the form of conveyors and forklifts have facilitated the process.

B. Transfer. Incoming items must now be transferred into the warehouse for storage in a designated location. There are multiple transfers over time since the stored items must also be transferred from the storage location to the shipping point. Additional transfers of the item may be required to facilitate access to other items in the warehouse.

C. Selection. Selection is the accumulation of a customized assortment of items to meet the needs of customers. The receipt of a customer order identifies the required items, which are then selected from warehouse stock and combined into an order.

D. Shipment. The final movement activity is shipment of the selected order. This includes verification of the item selection and loading of the order aboard the transport vehicle.

Location Planning

Warehousing is important to the firm since it improves service and reduces cost. Improvements in service are gained through rapid response to customer requests (time utility), which is a primary factor leading to increased sales. Improved customer service is achieved through the *production-positioned warehouse* (located near production facilities), which acts as a collection point for products needed in filling customer orders. For example, a firm may have several manufacturing plants, each producing a different product, located throughout the United States. A production orientation would locate a warehouse at each manufacturing plant, with each warehouse storing products from all plants. This system offers the advantage of providing superior service to the total product assortment.

An alternative choice would locate the warehouse near the point of product consumption (market positioning). Such market-positioned warehouses reduce costs by providing place utility. A *market-positioned warehouse* functions as a collection point for the products of distant firms with the resulting accumulation of products serving as the supply source for retail inventory replenishment. This approach allows large and cost-effective shipments from the manufacturer with lower-cost, local transportation providing service to individual retailers. Market-positioned warehouses may be owned either by the firm or the retailer (private warehouses), or they may be an independent business providing warehouse service for profit (a public warehouse).

A third choice is to locate the warehouse at some intermediate location between the production facility and product markets. The *intermediate-positioned warehouse* is considered on the same basis as the production-oriented facility—improved service. A firm may have several plants throughout the country, with each facility located, for economic reasons, near the source of raw materials. Under these conditions, the cost-effective warehouse location may be at some intermediate point.

Economic Justification

The basic consideration faced by the decision maker who must evaluate warehouse facilities is *cost*. Is there a cost advantage to be gained through the addition of a warehouse? If so, the facility is justified. It should be noted that facilities designed to improve customer service may add to the total logistic cost. Improved service, however, is expected to increase sales, which, in turn, reduces logistic costs as a percentage of sales. Improved service that does not increase sales offers no differential advantage and should be discontinued.

Recall that one function of the warehouse is to reduce transportation cost by permitting more economical full-load or truckload (TL) shipments between the firm and the storage facility. Distribution of smaller increments from the warehouse to the customer are made by local carriers.

For a warehouse to be justified, the cost of direct shipment to customers must be greater than the combined costs of truckload (TL) shipments, warehouse costs, and local delivery. For example, assume that the average sale is three units of a product and the less-than-truckload (LTL) transportation costs are $5.18 per unit. Each direct factory shipment incurs $5.18 per unit multiplied by three, or $15.54 in transportation costs per sale.

Assume that a warehouse facility may be acquired (through construction, purchase, lease, or rental) with storage space for 35,000 units at an average unit cost of $1.46. Further assume that TL shipments average $1.38 per unit and local delivery within the marketing area averages $1.52 per unit. Costs per unit, with TL shipments and the warehouse, are as follows:

$$\begin{aligned}\text{TL transportation charges} &= \$1.38 \text{ per unit}\\ \text{Warehouse charges} &= \$1.46 \text{ per unit}\\ \underline{\text{Local delivery charges}} &= \underline{\$1.52 \text{ per unit}}\\ \text{Total Transportation Cost} &= \$4.36 \text{ per unit}\end{aligned}$$

This equals an equivalent per-unit charge of $4.36 or $13.08 in transportation charges for an average sale of three units. The use of the warehouse is thus justified, as it yields a net cost advantage per sale of $15.54 $-$ $13.08 or $2.46.

Personnel Requirements

Excluding limited steps toward mechanization, such as conveyor belts or forklifts for movement and lifting, the warehouse requires a very labor-intensive activity. Additionally, many warehouses are faced with seasonal variations such as the large increase in stock that occurs prior to the Christmas holiday. These variations in activity require comparable variations in the work force if the facility is to meet cost objectives. Fortunately, it is relatively easy to determine the number of people required during any period of warehouse activity. The method used is similar to that employed in determining personnel requirements for the maintenance facility.

First, given the nature of the product stored in the warehouse, it is possible to accurately determine the timing and probable magnitude of periodic variations. For example, toys will peak during the months before Christmas, whereas candy will see an increased demand on Saint Valentine's Day.

Average stock levels and information regarding seasonal variations, coupled with expected demand for the product(s), permit a determination of the weekly (or any other selected interval) warehouse activity. Both receipts and outgoing shipments must be included in activity estimates. Given these data, personnel requirements are calculated as follows:

A. The average number of incoming shipments per week are multiplied by the time required for in-processing and storage of one shipment of the received material. The result represents total input processing time for one week.
B. Units in stock are divided by the average order size to determine the number of expected orders.
C. The number of expected orders is multiplied by the average processing time per order to determine total weekly out processing time.
D. The sum of weekly in- and out-processing times yields the total time required for warehouse activity.
E. The total time is divided by the work hours available (8 hours per day × 5 days per week or 40 hours) and equals the number of individuals required to perform warehouse activities for that week.

Here, as in the estimations of personnel requirements for maintenance, it is wise to remember that the employee who is on the job for 8 hours a day seldom works 8 hours. A more realistic estimate is attained if the work day is considered as either 7 or 7.5 hours (a workweek of 35 or 37.5 hours).

Material Handling

An overriding characteristic of the logistic channel is the need for a smooth and orderly progression of material flowing into the firm and products flowing to the customers. A primary function of the warehouse is to increase the efficiency of this flow by providing time and space utility. The ideal firm, however, would have no need for warehouses as the materials of production would arrive as needed and the products of production would flow from the firm in perfect synchronization with demand.

The utopian world is, unfortunately, beyond our grasp as of yet, and warehouses survive to offer protection against imperfect predictions of future demand. Warehouses, by their very nature, represent a breakdown in the flow of material through the logistic channel. The flow is interrupted as items are stored in the warehouse for subsequent reshipment. This interruption of the flow dictates material handling as a necessary part of the transportation process. Items must be removed from the transport vehicle and transferred to a designated location for storage. At a later time, the items must be transferred from the storage location and reloaded aboard the transport vehicle. Material handling is necessary; however, it should be minimized.

Material handling is, in some ways, an anomaly, in that the need for it arises from imperfections in the distribution channel. Any material handling represents an added cost; therefore, the best that can be expected is to minimize material handling, thereby adding the least cost to the distribution channel. With this in mind, it can be said that material handling has the objective of reducing total distribution costs.

Steps in the attainment of this objective are summarized as follows.

A. The optimum utilization of available space. The space required by a quantity of material varies significantly with the efficiency accorded placement and storage. It would not be unreasonable to attribute an apparent need for new warehouse facilities to "sloppy" and inefficient placement and storage practices. Another disadvantage can be delays in filling orders because of inefficiency.

B. The reduction of damage and waste. Herein lies a major opportunity for cost reduction. Damage and waste are direct additions to the cost of doing business, and reductions in this area have immediate impact.

C. Improvements in efficiency. Improvements in efficiency can also greatly reduce costs. Economy of movement in transferring material into and out of storage areas decreases the average time per shipment, thereby increasing productivity. Increased efficiency demands exact knowledge of the location and placement of each item. Automation and the ever-growing use of computers have contributed to significant progress in this area. However, excessive reliance on computers may lead to a shipment waiting on the dock for hours because "If it isn't in the computer, it doesn't exist!"

Major strides in the area of increased efficiency must, unfortunately, depend upon the redesign of existing facilities or the construction of new facilities specifically designed to enhance material-handling capabilities. Either alternative requires an extensive commitment of capital resources, and warehouses are typically low on priority lists. Nevertheless, improved efficiency must be rigorously pursued, because each small gain represents an improvement in the distribution pipeline.

> D. Improved working conditions. Material handling is the most labor-intensive logistic activity. This reliance on labor renders material handling overly susceptible to productivity changes in the individual employee. A clean, well-lit, efficient, and safe facility greatly increases productivity and, correspondingly, decreases costs.

Material handling, as one of the logistic activities, is concerned with the reception, transfer, selection, and shipment of material. These four types of handling are critical to the distribution pipeline and serve a single, defined objective: to keep the freight moving.

SUMMARY

Logistic facilities encompass a variety of structures established to meet the diverse needs of the elements comprising the logistic function. Examples of these facilities range from the training classroom to the supply room that maintains an inventory of spare parts.

The maintenance facility that supports production equipment and the products of production is a major component in this arena. Maintenance facilities are categorized by levels of maintenance, and a facility may support a single repair level or multiple levels.

Maintenance and repair are labor-intensive; therefore, personnel requirements are the primary factor in determining facility size. Personnel requirements are derived through analysis of the expected workload. Workload is, in turn, a function of the product being supported, the population of that product, and the MTBF and the average time required to perform each repair action. The combined total time for all expected repair actions represents a block of time that is readily converted into the number of personnel required. This analysis is equally valuable to the manager interested in setting up a new facility and one who wishes to evaluate the adequacy of current staffing.

The warehouse represents another major logistic facility. Warehouses, by providing time and place utility, reduce overall distribution costs and improve service to the customer. A warehouse may be

A. Production-positioned (located near a manufacturing facility). Production-oriented warehouses are justified on the basis of improved customer service.
B. Market-positioned (located near the point of product consumption). Market-oriented warehouses are justified on the basis of reduced cost.
C. Intermediate-positioned (located at some intermediate point between the manufacturing facility and the consumer). Intermediate-positioned warehouses are also justified on the basis of improved service.

Warehouses reduce total transportation costs by permitting firms to ship truckload (TL) or carload (CL) lots, thereby achieving more economical per-unit cost. A warehouse is justified if the full-load shipment cost, warehouse costs, and local delivery charges equal a lower per-unit cost than alternative small shipment rates. In this capacity, the warehouse receives large shipments and breaks them down into smaller increments for local delivery.

Warehousing involves the dual functions of movement and storage. Movement is further broken down into the material-handling functions of reception, transfer, selection, and shipment.

QUESTIONS FOR REVIEW

1. Identify two logistic facilities in addition to the maintenance and warehouse facility. Justify their inclusion as logistic facilities.
2. What single factor exerts the most influence on the size of a maintenance facility? What are some of the related factors that may influence size?
3. A product operates 10 hours per day, 5 days per week and has an MTBF of 2,200 hours and a population of 4,500 units. What is the expected number of intermediate-level repair actions per year?
4. What is the failure rate of the product in Question 3?
5. Assuming an average repair time of 6 hours, how many employees should the depot maintenance staff in Question 3 have?
6. Define MTBMA. Contrast it with the MTBCM.
7. Define the concept of time and place utility in relation to the warehouse.
8. Describe the process by which the warehouse reduces overall transportation costs.
9. How does the intermediate-positioned warehouse improve service to the customer?
10. Describe a method of estimating warehouse personnel requirements.

7

THE LOGISTICS OF RISK

Inventory is the logistics of risk, and decision making within this arena is one of the major problems facing the ILS manager. Decisions based upon reliability and maintainability analyses and informed judgment lead to commitments for particular inventory assortments. Correct decisions result in an inventory incorporating the right quantity of the right item, so that inventory assets are both needed and available when needed.

Erroneous decisions, on the other hand, result in improper inventory assortments. Improper assortments arise from a failure to include the correct item (inventory shortage) and the inclusion of too many items (inventory surplus). Inventory shortages may lead to extended shutdowns of the production line, marketing difficulties, and strained customer relations. They create disruptions in manufacturing and marketing operations. The answer is not, however, to include more in inventory, since the overstocked inventory is also a problem. Overstocking increases cost and detracts from profit through capital tied up in unneeded stock, added insurance costs, increased taxes, deterioration of items with a short shelf life, and the need for additional storage space. Severe overstocking could potentially require new and unnecessary warehouse facilities.

Logistics and inventory management are faced with the dual problem of what to include in inventory and how to achieve a balance between too much and too little. Unfortunately, the nature of inventory requires a decision today that is based upon future events. Thus, the task occurs within a planning period that is characterized by risk and uncertainty.

What Is Inventory?

In the simplest sense of the word, *inventory* is the accumulation of an assortment of items today for the purpose of providing protection against what may occur tomorrow. It is insurance. An ideal system would eliminate the need for inventory. Products would be made to order, and the finished good would be immediately transferred to the consumer, whose recognition of need corresponded with product completion. The materials of production would, in the same manner, arrive at the firm when needed in the production process. In this ideal system products would not be subject to breakdowns and the need for replacement items would coincide with product demand.

This zero inventory concept is contingent upon perfect correlation between supply and demand and is, as yet, an unattainable goal. Products do fail, and unforeseen perturbations in demand impact an equally variable supply. Inventories are necessary, and logistics must assure that each dollar invested in inventory is committed to the achievement of a specific objective.

Inventory is a major area of asset deployment, and every effort must be exerted to obtain an acceptable rate of return on this investment. The return on inventory is extremely difficult to measure, however, because the typical corporate profit and loss statement does not properly depict either the true cost of inventory or the inventory contribution to profit.

Forms of Inventory

An inventory is maintained to increase profitability through manufacturing and marketing support. Manufacturing support is provided through two forms or types of inventory: an inventory of the materials for production and an inventory of spare and repair parts for maintaining production equipment.

The materials inventory consists of the raw materials and purchased items that, through the transformation process, are converted into the finished good. Maintaining an inventory assures availability of these materials, thereby protecting against an interruption of production and avoiding a stock-out. This is economic justification for the establishment of a materials inventory. The results of a halt in production are easily measured and, if due to a stock-out, must be borne by inventory as the lack of required materials is a failure of inventory to meet demand. Stocks of material maintained in inventory must be sufficient to meet the demand derived from the production schedule.

The spare and repair parts inventory facilitates repair of production equipment. Malfunctions and breakdowns occur, and ready availability of the correct spare or repair part minimizes the time it takes to restore defective items to an operable status. Here again, it is relatively easy to determine the value of inventory. Added cost and out-of-service times engendered by a stock-out must be charged to the inventory function.

Inventory support to marketing is similar to that described for manufacturing. Marketing inventories consist of accumulations of the finished product (comparable to the materials inventory in manufacturing) and inventories of spare and repair parts that support the product (comparable to parts inventories supporting production equipment). Product inventories are maintained to assure availability upon demand, thus creating customer satisfaction. The customer may have spent days or even weeks before reaching a decision to purchase a product. Once the decision is made, however, a demand is created that must be satisfied. A lack of product availability is not acceptable, and the inevitable result is customer dissatisfaction.

The spare and repair parts inventory provides support to the product following ownership transfer. The firm maintains this inventory to assure availability when (1) the customer requests a part for repair of the product or (2) the customer returns the product to the firm for repair.

The forms of inventory have been separated into various types, but such divisions are largely artificial. A stock-out, for example, leads to similar results in each case.

A. A stock-out in the material inventory halts the production process through lack of the material necessary for production.
B. A stock-out in the spare and repair parts inventory halts the production process through the lack of items needed for the repair of production equipment.
C. A stock-out in the product inventory halts the marketing process through a lack of products for sale.
D. A stock-out in the spare and repaiur parts inventory halts the marketing process through a lack of parts to either sell or effect repair of the product.

Other similarities exist. For example, each form of inventory must accumulate an assortment of items in anticipation of future demand, and they share equal concern over stockage levels. Answers to the basic inventory questions of what to stock and how much to stock are derived in the same manner regardless of the ultimate objectives of specific inventory types. For this reason, the remainder of this chapter discusses inventory without reference to specific types.

A Conflict of Functions

Previous examples indicated the relative ease of determining the cost associated with the lack of needed items in inventory. The problem becomes much more complex when attempts are made to specify the desired inventory level. How much is enough cannot be definitively answered, given the current sophistication of measurement techniques.

The inability to answer this question creates conflict within the firm concerning the appropriate level of inventory commitment and allocation of corporate resources. The financial function, for example, does not want to tie up today's cash in nonliquid assets such as a large inventory. The overriding tendency is to minimize inventories, thereby improving cash flow.

Manufacturing, on the other hand, prefers to stockpile large quantities of production materials for protection against disruptions in their supply. In a similar manner, marketing desires large and diverse inventories of the finished good to guard against stock-outs and to assure product availability.

The Functions of Inventory

As previously stated, the basic (and only) reason for maintaining an inventory is to reduce costs and improve profitability. The successful attainment of this objective depends, in part, upon four distinct functions served by inventory: geographic specialization, decoupling, balancing supply and demand, and safety stock.

The first of these functions, *geographic specialization,* permits economically advantageous separation of individual operating units within the firm. The manufacturing function of the firm, for example, may consist of several operating units with each producing a specific part or component of the finished product. Factors of production such as access to a source of raw materials and the availability of power or labor may dictate economical manufacturing locations that separate the operating units and move them a considerable distance away from the marketing area. Geographic specialization permits production where it is most economical. Then, through internal inventory transfer, the various parts and components are assembled into the final product. This same principle is employed when a warehouse creates an assortment of goods by collecting products from various locations, thereby offering customers the opportunity to purchase a product mix.

The second function, *decoupling,* is the equivalent of geographic specialization applied to a single location. It is designed to provide maximum efficiency of operations within the firm through the use of inventory as a buffer between the production and marketing functions. Decoupling begins with the stockpiling of materials and work in process within the manufacturing complex. This permits economies of production through the elimination of work stoppages because of stock-outs. Additionally, products are produced to inventory, which leads to economical production runs and a decoupling of the manufacturing task from shifts in demand.

On the marketing side of the firm, a warehouse inventory produced in advance of need acts as a buffer between manufacturing and production and permits stockpiling of the finished product. The resulting delay in product distribution leads to large quantity shipments at minimum freight charges per unit.

Inventory also acts to achieve a *balance between supply and demand.* This is most apparent in seasonal products or products having seasonal production and a year-round demand. A *seasonal product* is one that is characterized by low demand throughout most of the year with a high demand period centered around an event or climatic condition. Bathing suits, lawn furniture, antifreeze, and winter clothing are all examples of seasonal products. Wine and orange juice are excellent examples of products having seasonal production and year-round demand. Production, in both instances, takes place during a certain season of the year with a relatively stable demand throughout the year.

Success in maintaining a balance between supply and demand requires regularity in the demand cycle. A knowledge that the demand for winter clothing, for example, will increase dramatically in September and October permits the manufacturer to schedule production to assure product availability.

The final function of inventory is as a *safety stock*. Safety stock is inventory that has been set aside to protect against short-term variations in either demand or supply. It is a direct result of the risk and uncertainty that characterizes inventory planning. Safety stock provides protection against two types of uncertainty, type 1 and type 2. Type 1 uncertainty is related to demand that exceeds projection. As an example, assume that an inventory has been established to meet a projected demand for fifty units of a product. What happens if the demand is for fifty-two units? In the absence of safety stock (an increment of inventory exceeding projected demand), the sale of the additional two units will be lost because of a stock-out.

Type 2 uncertainty is related to delays in the replenishment cycle. In this instance, safety stock items are maintained to protect against unexpected extensions of the delivery schedule.

Inventory Risk

Establishment of an inventory entails a considerable degree of risk to the logistics practitioner. Both inventory content and quantity are based upon judgment and a variety of forecasting techniques. Events that fail to materialize and those that exceed expectations represent problems to inventory management. The risk, however, is not uniform across the logistic channel. The nature and depth of risk vary with the type of firm.

Manufacturing Risk

Inventory risk for the manufacturer begins with the commitment to acquire raw materials, parts, and components for the production of a product. It continues through work-in-process and the finished good, ending only with satisfactory sales of the product. The decision to manufacture a product requires a commitment of resources well in advance of possible sales. This relatively long time duration is characteristic of manufacturing risk and, given the changing nature of the consuming public, increases the degree of risk to the manufacturer. The product may not find a receptive audience upon reaching the market, and the product that does not sell leaves the manufacturer with large, expensive inventories of raw materials; work in process; and finished products.

The manufacturer, in comparison to the wholesaler or the retail merchant, has a very narrow line of products. Because of this, manufacturing inventory risk has a relatively deep and narrow base, as the failure of a single product may have a severe impact on profitability. This depth is another characteristic of manufacturing risk.

Wholesale Risk

The wholesale merchant receives economic justification by purchasing from the manufacturer in large quantities and selling to the retailer in smaller quantities. Typically, several manufacturing firms supply products to the wholesaler, resulting in a product mix at the wholesale level. This mix spreads inventory risk over a much wider product base than that of the manufacturer. Wholesale inventory risk, therefore, has a wider base than that of the manufacturer, since the failure of a single product has a less severe impact.

The wholesaler is, however, faced with inventory risk of long time durations. Consider the product having a relatively short production period and year-round demand. The wholesaler must, in this case, invest in a large inventory during production in order to meet demand throughout the year. The long period of time between product acquisition and sale increases wholesale risk.

Retail Risk

Retail risk has a wider base of less depth and of shorter duration than that of the wholesaler or the manufacturer. The typical retailer stocks thousands of items; therefore, the failure of a single product has relatively little impact on the firm. However, specialty shops that offer a limited range of products have an inventory risk whose base is narrower and deeper than that of the mass merchandiser.

The Inventory Cycle

A small number of firms such as drop shippers place orders for merchandise only after sales have been made; therefore, they require no inventory. Most firms, however, require an inventory in order to operate effectively. Given that inventories are necessary, the problem now becomes one of determining (1) what to stock in inventory, (2) when to take action in ordering or reordering inventory, and (3) in what quantity should the components of inventory be ordered? Offering practical and cost-effective answers to these questions constitutes the most difficult area of inventory management.

The first question, what to stock in inventory, is partially answered by virtue of the specific business opportunity selected by the firm. For example,

> A. The manufacturing firm is established to produce a product. The choice of products determines the materials inventory, the work-in-process inventory, and the finished goods inventory. The equipment of production, also established by the choice of product, determines the repair parts inventory.

B. The retail merchant and the wholesaler choose to operate within a specific class of product or products. This choice again determines the components of inventory.
C. The repair facility, whether a manufacturing component, a retail component, or an independent enterprise, exists to support a type or class of product. Reliability analysis identifies what parts are subject to malfunctions, thereby specifying inventory content.

The remaining questions, when to order or reorder and in what quantity, are much more difficult to answer and require an understanding of average inventory.

Average Inventory

Average inventory is a measure of inventory contents over time. The concept of average inventory is best illustrated by the simplistic financial example wherein an individual is loaned $100 on the first day of the year, to be paid back at regular intervals throughout the year. As the money in hand is $100 at the beginning of the year and $0 at the end of the year, the average amount of money held during the year is $50. Now, to carry this analogy one step further, assume that the individual held a constant amount of money equal to $10 throughout the entire year. The average money in hand now becomes $50 (from the loan) plus the $10 in constant funds.

Average inventory is equivalent to the second part of this example with the $50 representing base stock and the $10 representing safety stock. *Base stock* is defined as the portion of inventory resulting from the reorder process. Base stock and *safety stock,* which provides protection against type 1 and type 2 uncertainty, equal *average stock.*

Pipeline Inventory

Replenishment inventory items are ordered through the logistic channel with subsequent shipment from the supplier to the buyer. There is a finite and highly variable interval of time associated with the transportation activities interconnecting the supplier with the firm. During this time the items have been removed from physical control of either party, yet, one or the other must maintain ownership. The party (buyer or seller) maintaining ownership also retains these items as a part of moving inventory. This portion of total inventory is referred to as the *pipeline inventory.*

Ownership of pipeline inventory is based upon the conditions of sale. If purchase is f.o.b. destination, ownership is retained by the seller until delivery; pipeline inventory is owned by the selling party. The opposite condition exists when the purchase is f.o.b. origin. In this instance, ownership is transferred to the buyer when the items enter the logistic channel. Items in the logistic pipeline are owned by the buyer and must be counted as part of average inventory.

Inventory Performance Cycles

The establishment of an inventory is (assuming proper selection of inventory items), followed by gradual consumption of those items. Consumption will, in time, deplete inventory to the point where replenishment is necessary. The interval between a fully stocked inventory and the time it is again fully stocked through replenishment is the *inventory performance cycle*. The inventory performance cycle is different for each item in inventory, as consumption rates vary between items. Note that the performance cycle incorporates transportation time, since replenishment does not occur until the items have been physically received by the firm.

Let us consider a brief example, under specific and assumed conditions, to clarify the concept of the inventory performance cycle. These assumed conditions are as follows:

 A. The duration of the performance cycle is constant.
 B. The acquisition time (lead time or the time between the request for a replenishment item and its receipt) is constant.
 C. The consumption rate per day is constant.

Although these assumptions ignore the complexities introduced by uncertainty, their use does not invalidate a simplified illustration of the inventory performance cycle.

The saw-toothed diagram in Figure 7.1 illustrates the inventory performance cycle. The assumed constants are (1) a 20-day inventory performance cycle, (2) a lead time of 10 days, and (3) a consumption rate of 2 units per day. Figure 7.1 identifies the beginning (day 0) inventory as 40 units. Consumption, at the rate of 2 units per day, would reduce this inventory to 0 units at the end of 20 days. The beginning inventory of 40 units, decreasing to 0 units 20 days later, results in an average inventory of 20 units. (The stock on hand exceeds 20 units one-half of the time and is less than 20 units for the other half of the time.) Average inventory is equal to base inventory as safety stock is 0.

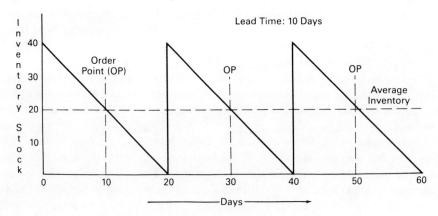

Fig. 7.1. The Inventory Performance Cycle

The dollar value of inventory is a function of average inventory times the unit price of each item in inventory. Therefore, assuming a unit price of $15.50, the value of this inventory is $310.00 ($15.50 per unit × 20 units).

This method of establishing inventory value is oversimplified, since it ignores the value added to inventory stock as a function of order processing, transportation charges, handling storage, and other related costs. These costs are discussed in a later section.

The initial assumptions eliminated uncertainty; therefore, a shipment of replenishment items may be scheduled to arrive on day 20 (the same day that inventory quantities reach 0). The new shipment restores inventory to 40 units, and consumption now continues through the next inventory cycle. Replenishment items have a lead time of 10 days; therefore, if the new shipment is to arrive when inventory reaches 0, the order for replenishment stock must be placed 10 days prior to this time. This order point (OP) takes place when 20 units are on hand and corresponds to the 10-day lead time required for replacement orders. Thereafter, an order is placed whenever on-hand inventory decreases to 20 units (the order point).

In this example, orders are placed every 20 days or twelve times per year for an assumed work year of 240 days. The 12 orders of 40 replenishment units per order (480 units) are the replacements for inventory stock consumed throughout the year. This indicates that the inventory will turn over twelve times. It is apparent that inventory turnover refers to the consumption of inventory stock, which may consist of (1) the use of raw materials or parts in the production process, (2) the use of components or kits of parts in completing work in process, (3) the sale of completed products, or (4) the use of spare or repair parts for maintenance.

Figure 7.2 expands on the previous example by including provisions for 5 units of safety stock. The average inventory is now 25 units, comprised of 20 units of base stock plus the 5 units of safety stock. The value of this inventory reflects a concurrent increase to $387.50. Also, the order point has increased as orders must now be placed when inventory stock reaches 25 units.

It is, of course, possible to place orders for replenishment stock at other than the 20 days illustrated in Figures 7.1 and 7.2. Changing the order cycle, however, impacts both the inventory performance cycle and average inventory. Figure 7.3 illustrates this approach by depicting a 40-day inventory performance cycle.

Placing orders less frequently, with a constant consumption rate, means that each order must be for a larger quantity of inventory stock. The converse is also true in that inventory stock must be larger, as there is a longer period of time before the new order is received by the firm. These conditions are depicted in Figure 7.3. Inventory has been increased to 80 units, which, at the previously assumed consumption rate of 2 units per day, will reduce on-hand stock to 20 units on day 30. A replenishment order for 80 units is placed at this time as the indicated lead time of 10 days indicates that the order will be received coincident with on-hand stock reaching 0. Inventory has been restored to 80 units, and the cycle begins anew. The inventory thus turns over six times in a 240-day work year as replenishment

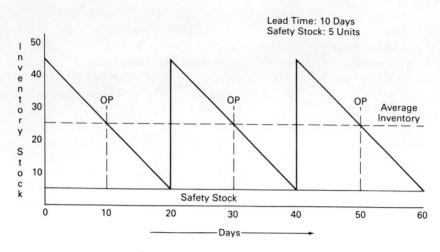

Fig. 7.2. The Addition of Safety Stock

occurs every 40 days. In addition, the average inventory has increased to 50 units with a value of $775. Substantially more cash assets are tied up in this larger inventory.

Notice that safety stock has been raised to 10 units with the increase in length of the inventory performance cycle. This is in line with the greater degree of inventory risk that is encountered with an increase in the time interval between replenishment periods.

Figure 7.4 illustrates the effect of an increased inventory cycle frequency. In the figure, average inventory has been reduced to 12 units and orders are placed every 10 days (inventory turnover of 24).

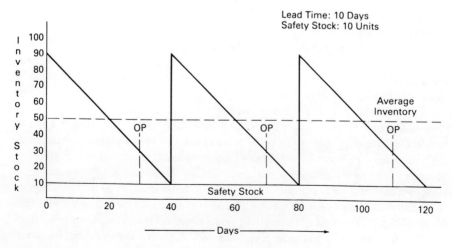

Fig. 7.3. A Forty-Day Performance Cycle

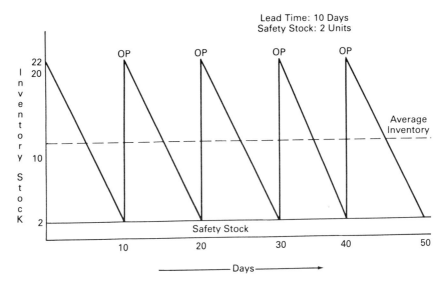

Fig. 7.4. A Ten-Day Performance Cycle

Adding Pipeline Inventory

The preceding examples did not consider the effect of transportation times on replenishment stock that is purchased f.o.b. origin. This stock becomes the property of the buyer upon entering the pipeline and must be included as part of inventory. The effects of this pipeline inventory may be easily demonstrated by referring to Figure 7.2. An order is placed for 20 units 10 days before on-hand inventory reaches 0. The order immediately enters the pipeline for transport to the firm, arriving 10 days later. The 20 units in the pipeline represent an average pipeline inventory of 10 units, which must be added included with inventory. This results in a true average inventory of 35 units (20 units of base inventory plus 5 units of safety stock plus 10 units of pipeline inventory).

To amplify on the effects of pipeline inventory, consider a geographically separated operation wherein the firm purchases raw materials from a remotely located supplier (Figure 7.5).

Pipeline time between the firm and its supplier is 27 days. The firm places orders every third day for 500 units at a cost of $375 per unit. Therefore, at any given moment there are nine individual orders of 500 units each in the pipeline or a total of 4,500 units. The average pipeline inventory of 2,250 units has a value of $843,750.

The firm now locates an alternate and much closer source of raw materials with a pipeline time of 6 days. Orders are still placed every third day for 500 units per order. Now, however, only two orders (1,000 units) are in the pipeline at any one time, resulting in an average pipeline inventory of 500 units. The value of this

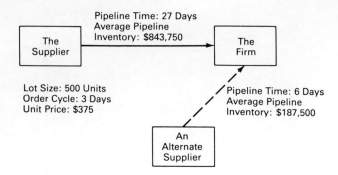

Fig. 7.5. The Effects of Pipeline Inventory

inventory is $187,500 or $656,250 less than the 27-day pipeline. The extra money may now be used for other functions within the firm. A comparable effect could have been achieved with a reduction in transportation time from the original supplier.

The effects of pipeline inventory may also be used to assist in the economic justification of warehouse facilities. To illustrate this, assume that the warehouse in Figure 7.6 supplies two markets with the identical product. The transportation time to markets A and B are 5 days and 12 days, respectively. One unit of the product has a value of $200 and is shipped to each market daily. At any given time, 5 units of the product are in the pipeline to market A and 12 units are in the pipeline to market B. This results in a combined average pipeline inventory of 8.5 units having a value of $1,700.

The wholesale merchant decides to acquire another warehouse facility located near market A, thereby reducing transportation time to 4 days (Figure 7.7). Under the previously identified conditions, the pipeline inventory is now 5 units to market B and 4 units to market A. This gives a combined average pipeline inventory of 4.5 units having a value of $900, a reduction of $800 in committed inventory funds. Given the capacity of common carriers, the combined savings resulting from multiproduct shipments may definitely justify additional warehouse facilities.

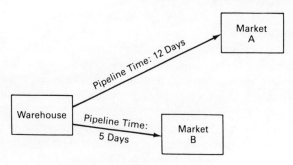

Fig. 7.6 A One-Warehouse Pipeline

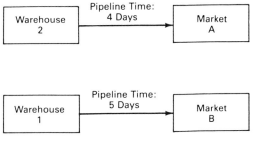

Fig. 7.7. A Two-Warehouse Pipeline

Elements of Inventory Cost

The initial acquisition cost of inventory items represents only one part of the cost of inventory. Acquisition costs must be supplemented by a variety of other factors; each of which acts to increase the effective cost of items in inventory. Two of these, taxes and insurance, are relatively easy to isolate and identify. *Taxes* are a direct levy against declared inventory value, and insurance costs are a payment based upon value and risk over time. Other costs, as discussed in succeeding paragraphs, are much harder to quantify.

Transportation Costs

At first glance transportation costs appear to be relatively easy to quantify. Charges are, after all, readily identifiable from financial records and should represent a simple addition to the purchase cost. This would be true if the shipment consisted of only one item of inventory stock. Multi-item loads consisting of many different inventory items in a variety of sizes and shapes require considerable analysis to properly allocate costs to individual elements. The problem is compounded when shipments include items or products in addition to inventory stock.

Storage Costs

Storage costs must be allocated to individual items, since they do not contribute directly to inventory value. Public warehouse charges are somewhat easier to calculate since financial data are readily available. This cost can then be allocated among inventory items on a per-unit or cubic-foot basis.

The privately owned warehouse facility must base these computations on total annual depreciation expense. This expense is then allocated to inventory items on a per-unit or per-cubic-foot basis as in the public warehouse. Additionally, the private warehouse must allocate a pro rata share of fixed and variable costs to inventory items. In either case, care must be taken to make appropriate allowances for space that is not being used.

Obsolescence Costs

Obsolescence costs are associated with inventory items that deteriorate while in storage and are not covered by insurance. They can also include losses when the item becomes obsolete through the introduction of a new model or design. Obsolescence costs should be approached with caution and allocated on a per-unit basis.

Opportunity Costs

Opportunity costs derive from the fact that money invested in inventory loses its earning power, reduces capital availability, and is not available for alternate uses. The appropriate charge to place on capital invested in inventory remains controversial; however, logic supports the use of the prime interest rate. The reasoning for this is that the monies invested in inventory can be replaced through borrowing at that approximate rate. This type of opportunity cost is also referred to as the cost of capital or the cost of money.

Opportunity costs take on a greater significance in the case of unused inventory or inventory that is larger than needed. These costs are related to possible alternative uses for the space occupied by the unnecessary inventory. As may be imagined, these costs are extremely difficult to quantify.

Order Costs

Considerable costs are generated as the result of order placement, and order processing costs are a significant part of inventory valuation. These costs include the research necessary to locate acceptable suppliers; to prepare, communicate, follow up, and update orders; to maintain records; and to supervise employees. Order costs are developed for each activity element and then added together to obtain the total cost of placing orders. The total figure must include assigned shares of both fixed and variable costs associated with this example. The total cost is then divided by the number of orders anticipated for the year. For example, if the calculated fixed cost associated with order placement is $450,000.00 per year and the number of expected orders is 300,000, the fixed cost per order is $1.50. Assumed variable costs of $600,000.00 would add $2.00 to this amount, resulting in a total cost per order of $3.50.

Substantial differences exist between firms concerning the actual cost associated with placing an order. The important thing is to assess as accurately as possible all elements of applicable fixed and variable costs before assigning them to order processing activities. Once the per-order cost has been established, this assumption is held constant regardless of the number of orders actually filled. This assumption of linearity is incorrect, however, as the actual cost per order varies above and below the estimated amount. The aggregate error over time is small.

Production and Inventory Control

Production and inventory control is a merging of two disciplines for the expressed purpose of improving the operations of each. Inventory control, in the early stages of this activity, consisted of monitoring inventory levels, issuing purchase orders at or near safety stock levels, and assigning due or suspense dates for the requested items. In many, if not the majority, of cases, due dates turned out to have little meaning and many of the requested items were late. The requested items were necessary for continued production; thus, in many firms the production control function consisted of expediting. No one asked why inventory control could not establish realistic due dates to support manufacturing needs. Production control continued trying to expedite overdue orders.

The advent of the computer provided a tool that seemed ideally suited to the task of production and inventory control. Initial attempts, however, proved very discouraging. The computer worked well in finance but could not seem to solve the problems of production and inventory control. The reasons behind this apparent contradiction were obvious. Accounting had a tried-and-true system that worked; therefore it worked equally well, and much faster, when run by a computer. Production and inventory control, on the other hand, did not have a system that worked. An unworkable system transferred to a computer remains unworkable.

Formal and Informal Systems

Production and inventory did possess formal procedures manuals that included detailed presentations of the steps necessary to implement a working system. Manuals abounded with instructions to place orders when on-hand stock reached the order point and to establish due dates based on lead times.

When due dates were not met, formal systems were replaced with informal systems that attempted to get the job done through hot lists and expediting. Hot lists are a listing of the overdue items that are most desperately needed. They provided the expediter with a priority sequence of suppliers to beg, plead, and cajole for speedy delivery of overdue items.

This circumstance led to two systems of production and inventory control in most companies: The formal or *push system,* in which an inventory control system released purchase orders to suppliers, and the informal *pull system* with expediters attempting to wrest overdue items from those same suppliers.

Functions of Production and Inventory Control

The task of production and inventory control is, quite simply, to generate plans that synchronize order processing activities with manufacturing activities, that people will accept, and that can be used to measure performance, thus permitting management of both production and inventories. In order to accomplish this objective, production and inventory control must fulfill the following basic functions:

A. Priority planning
B. Capacity planning
C. Priority control
D. Capacity control

Successfully meeting these four basic functions eliminates the artificial distinction between production control and inventory control.

Note by the sequence of the four basic functions that planning must precede control. In fact, it can be said that there can be no control if it is not preceded by realistic and valid planning. Note also that priority planning takes precedence over the other functions. If there is no priority planning, none of the other functions will work. Priority planning does, however, vary slightly among the different types of manufacturing firms. For example, manufacturing firms may be classified by the type of product that is produced, such as

A. One-piece products made to order.
B. One-piece products made to stock.
C. Assembled products made to order.
D. Assembled products made to stock.

Priorities are quite basic in the firm that makes a one-piece product to order. The buyer requests the product and accepts a delivery date. This priority does not change unless requested by the buyer.

In contrast, the firm making products to stock, one-piece or assembled, is faced with the potential problem of updating priorities as materials needed in the production process flow into the manufacturing facility. A delay in one material input may require that priorities be shifted to an alternate product pending receipt of the delayed item. This problem increases in significance as lead time increases.

Firms that produce an assembled product are faced wth additional complexities because dependent priorities come into play. For example, assume that an item requires a sequential installation of eight parts to make up the finished product. If the fourth part to be installed cannot be made available as scheduled, the product cannot be completed, and any existing priorities (parts 5 through 8 and the completed product) should be readjusted. It is patently foolish to continue attempting to expedite a part that is installed later in the production sequence when an earlier part delay has negated the production schedule. Yet, this is exactly what happens in informal systems because they are incapable of "tracking" the numerous items in a typical manufacturing operation.

The formal inventory systems did not work, yet they were transferred to the computer, where they still did not work. The informal system had to be replaced with a formal system that worked before the computer could become a tool of production and inventory control.

Economic Order Quantities

The introduction of computers into production and inventory control was characterized by a simple mechanization of previous practices. Production control, consisting largely of stock chasing, could not be readily adapted to the computer; thus, this area suffered little immediate impact. The essentially clerical function of inventory control, however, appeared to be an ideal candidate. This had the additional advantage of coinciding with the prevailing opinion that economic justification of computers was provided by the number of clerks they displaced.

This introductory phase viewed inventory control as being primarily responsible for a determination of two fundamental inventory issues: (1) How much should be ordered? (2) When should it be ordered? The first of these questions dealt directly with the lot size of individual orders. Orders in large quantities typically provide quantity discounts; however, there were offsetting factors such as the increased storage costs engendered by resulting large inventory. What is the most economical lot size when considering placement of an order? The concept of such an economic order quantity (EOQ) has existed for years and, in fact, "rediscovered" at regular intervals and presented as the ultimate solution to the problems of inventory. The formula for the economic order quantity is

$$EOQ = \sqrt{\frac{2AS}{I}} \qquad (7.1)$$

where: A = annual usage in dollars
S = setup (ordering) costs in dollars
I = inventory carrying cost as a percent of inventory value

The use of EOQ can best be illustrated through the assumed values as pro-

Table 7.1. SAMPLE EOQ INPUT FACTORS

Annual usage	3,000 units
Unit cost	$1.25
Inventory carrying cost	20%
Ordering cost	$15.00

vided in Table 7.1. Substituting the values listed in Table 7.1 in Eq. (7.1),

$$EOQ = \sqrt{\frac{(2)(3,000)(\$1.25)(\$15.00)}{0.2}}$$

$$= \sqrt{\frac{112,500}{0.2}}$$

$$= \sqrt{562,500}$$

$$= \$750.00 \text{ or } 600 \text{ units at a cost of } \$1.25 \text{ per unit}$$

This basic equation for determining economic order quantities has been refined, modified, and presented in a variety of forms with only the objective of determining the optimum order size remaining constant.

The EOQ concept is based upon the facts that total inventory costs are minimized at some definable purchase quantity. Total inventory costs are a function of the number of orders that are processed per unit time and the costs of maintaining an inventory over and above the cost of items included in inventory. These relationships are as follows:

A. Ordering costs are assumed to increase at a linear rate. If the cost is $15 for one order, it will be $150 for ten orders.
B. Inventory maintenance costs per item decrease as inventory quantity increases. A $10 maintenance cost for one item is reduced to $5 per item if two are stored in the same space.

This relationship is depicted graphically in Figure 7.8.

The concept of economic order quantity is not without faults, however. It ignores transportation costs, which may be significant for smaller shipments. The EOQ concept also ignores the effects of quantity or volume discounts and assumes linear usage with independent demand. None of these factors may be safely ignored, and their lack of consideration in EOQ calculations have contributed to its decreased significance. The question of when to order is now considered to be the most important of the two issues.

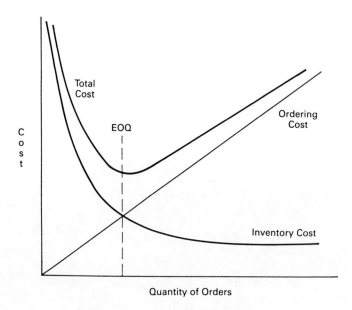

Fig. 7.8. The Economic Order Quantity

Order Point

The "when" of ordering was first answered through the use of order points. *Order points* incorporate lead time, demand, and safety stock into a relationship that reduces reordering to a simple mechanical process. The equation expressing this relationship is

$$OP = DL_t + SS \qquad (7.2)$$

where: OP is the order point
D is the demand
L_t is the lead time
SS is the safety stock

For example, assume a lead time of 6 weeks, a demand of 100 per week, and a safety stock of 250. The resulting order point is

$$OP = (100)(6) + 250$$
$$= 600 + 250$$
$$= 850 \text{ units}$$

Thus a replenishment order should be processed when the on-hand inventory decreases to 850 units. The quantity that should be ordered when on-hand stock decreases to this level is defined by the EOQ.

The OP developed with Eq. (7.2) can be used in conjunction with the EOQ to determine average inventory, which is equal to

$$\bar{I} = SS + \frac{Q}{2} \qquad (7.3)$$

where: $\bar{I}$ is the average inventory

SS is the safety stock

Q is the order quantity

Substituting the previously used values of 600 and 250 for order quantity and safety stock, respectively, average inventory becomes

$$\bar{I} = 250 + \frac{600}{2}$$
$$= 250 + 300$$
$$= 550 \text{ units}$$

This method of calculating the order point assumes an accurate accounting of each inventory item at all times (a *perpetual review*). Perpetual reviews are, however, only one method of tracking inventory. An alternate approach is to use a periodic review wherein inventories are checked at discrete intervals of time. The use of periodic reviews requires modifying the equation for calculating the order point as follows:

$$OP = D\left(L_t + \frac{P}{2}\right) + SS \tag{7.4}$$

where: D is the demand
L_t is the lead time
P is the review period
SS is the safety stock

Using a periodic review period of 4 weeks with the previously assumed values for perpetual review, the order point is

$$OP = 100\left(6 + \frac{4}{2}\right) + 250$$
$$= 100(6 + 2) + 250$$
$$= 100(8) + 250$$
$$= 800 + 250$$
$$= 1{,}050$$

Notice that the order is placed at a higher level of on-hand stock with periodic reviews. This is necessary since the on-hand inventory may reach safety stock levels shortly after the completion of a periodic review. The resulting delay before the next review may permit this inventory to fall below safety stock levels.

As with the EOQ, order points incorporate specific assumptions that limit their usefulness. First, the order point assumes that lead times can be accurately specified and are fixed. Second, it assumes that due dates do not change once they have been established. Neither assumption is valid in the real world. A process that meets the needs of production and inventory control must be able to time phase requirements with respect to need.

Discrete Lot Sizing

Inventory control situations rarely reflect the uniform usage rates that are characteristic of EOQ reorder situations. Manufacturing demands in particular normally occur at irregular intervals and for varied quantities because demand is dependent upon the production schedule. Parts and materials must be available to manufacture a given product. Between manufacturing intervals, or when manufacturing shifts to an alternate product, there is no need to maintain this inventory. This, of course, assumes that the requisite inventory can be obtained when needed.

This inventory is subject to manufacturing demand and requires a modified approach to determine order quantities. This modified approach is referred to as *discrete lot sizing.* Some of the more common types are (1) lot-for-lot, (2) period-order-quantity, and (3) time-series discrete lot sizing.

Lot-for-lot discrete lot sizing is the most basic form. It is a simple one-for-one approach that matches the order with demand over a specified interval of time. For instance, a projected demand for 7 units triggers an order for 7 replenishment units. No consideration is accorded quantity discounts, transportation charges, or order processing costs under lot-for-lot acquisitions.

The *period-order-quantity* (POQ) technique builds upon the EOQ to derive order placement intervals. It begins with the EOQ as calculated with Eq. (7.1). The EOQ is then divided into the projected demand to determine the number of orders that must be placed. This figure is then divided into either 52 or 12 to determine the weekly or monthly interval between orders.

To illustrate the period-order-quantity technique, consider the data expressed in Table 7.1 that yield an EOQ of 600. The POQ technique is as follows:

$$\text{Economic order quantity (EOQ)} = 600$$

$$\text{Projected demand} = 3{,}000$$

$$\text{Orders per year} = \frac{3{,}000}{600} = 5$$

$$\text{Order interval} = \frac{12}{5} = 2.4 \text{ months}$$

$$= \frac{52}{5} = 10.4 \text{ weeks}$$

Thus, under POQ, 5 orders are placed during the year, each order is for 600 units (the EOQ), and the orders are placed at intervals of approximately 2½ months.

Time-series lot sizing combines the requirements for several periods to develop a uniform order quantity. This technique is based upon a prior determination of order placement intervals. For example, assume that a decision has been made to place orders on a quarterly basis (four times per year). With a projected demand of 3,000 per year, four orders of 750 units are required.

Time-series lot sizing is more dynamic than EOQ in that the order quantity may be adjusted to conform to the latest and best estimates. Still, it fails to meet the requirements engendered by rapid and frequent changes in the level of demand. These requirements have been answered, however, by a process called *material requirements planning* or MRP. The subject of MRP and how it responds to changing priorities within the manufacturing arena is presented in Chapter 8.

SUMMARY

Inventory is properly referred to as the logistics of risk in that it entails a commitment of financial assets on the basis of projected events that may or may not occur. Too little inventory, the wrong type of inventory, and stock-outs adversely impact the firm's profits or good will. If there is too much inventory, the firm has too much committed capital, which would be better employed in other areas.

An inventory represents an accumulation of items that have been selected to support functions within the firm. The production function is supported by inventories of the materials of production, work-in-process, and spare and repair parts for maintaining the equipment of production. Marketing, on the other hand, is supported through inventories of the products of production and spare and repair parts supporting those products.

Inventory risk varies in both type and degree as a function of the area in which the firm chooses to operate. The manufacturing firm produces a small number of products in comparison to the retail merchant and therefore faces an inventory risk whose base is deep and narrow. Retail merchants who handle a diversity of products face a risk whose base is wide and shallow. The risk facing the wholesale merchant is somewhere between that faced by the manufacturer and that by the retailer.

Production and inventory control attempts to combine control of production with control of inventory. Formal procedures were established for this purpose; however, they addressed only what *should* happen. They were incapable of responding to changes. Consequently, informal systems arose to take their place. Informal systems were characterized by hot lists and the expediting of delayed items in an attempt to make the system work. This effectively reduced production control to the role of chasing stock.

Inventory control, on the other hand, was perceived primarily as a clerical function related to the release of purchase orders and the determination of due dates. This clerical aspect led to an early use of computers in inventory control. At first, computers merely addressed the issues of how much to order and when to order using established techniques. These techniques were primarily related to determinations of the economic order quantity (EOQ) and the order point (OP). EOQ answered the question of how much to order, and OP the question of when the order should be placed.

There are disadvantages to this approach: False assumptions are made regarding the demand, lead times, and due dates. Demand is assumed to be linear over time, lead times are assumed to be accurate, and due dates are not subject to change. This does not reflect the real world, where due dates are a function of demand and where demand changes regularly. Discrete lot sizing represents an attempt to overcome some of these difficulties; however, it too falls short of meeting the requirements of the manufacturing environment.

QUESTIONS FOR REVIEW

1. Why is inventory referred to as the logistics of risk?
2. Describe the impact of an inventory that lacks essential items of inventory stock.
3. Define the inventory function of decoupling.
4. Describe the inventory risk facing the manufacturer.
5. The shipping time is 30 days, the order cycle is 5 days, and each shipment consists of 100 units. What is the average pipeline inventory?
6. Using the data in Question 5, what is the effect on average inventory levels when replenishment orders are changed from every 5 days to every 10 days?
7. What is an economic order quantity?
8. Define perpetual review. How does it differ from periodic review?
9. What is the relationship between the order quantity and the average inventory?
10. What are the elements of inventory cost?

8

MATERIAL REQUIREMENTS PLANNING

Material requirements planning (MRP) is a set of logically related procedures, decision rules, and records specifically designed as a time-phased inventory control methodology. MRP incorporates inventory control, shop floor scheduling, and capacity planning into a management system that works. It is a powerful tool and one that has had a profound impact on the manufacturing firm. MRP has been so successful in practice that it has made previous approaches to inventory control obsolete. The success of MRP led to the evolutionary development of MRP II (manufacturing resource planning), which added the functions of finance, marketing, and purchasing to MRP. Counterparts to MRP, referred to as DRP (distribution requirements planning) and DRP II (distribution resource planning), have been developed in a similar manner to support physical distribution.

The unique feature of MRP is time phasing, the incorporation of time into previous forms of inventory control. Time phasing provides MRP with the capability of meeting three fundamental goals of the firm:

A. Reduction of inventory
B. Improvements in customer service
C. Optimized production

MRP: The Beginning

MRP began as another technique for determining when to place an order. It was, however, quite different from the OP (order point) approach. Order point systems assumed that each item in inventory is independent of all other items and can be ordered independently. MRP, on the other hand, looks upon items as part of an assembled product, treats their priorities as dependent, and calculates future demand as a function of the master production schedule (MPS). The MPS is a schedule of what is demanded and when it is demanded. It is the tool that is used to schedule the production necessary to meet those demands.

The MPS is, in itself, a forecasting tool developed through a combination of sales forecasts, warehouse stock, safety stock requirements, spare and repair parts requirements, and inventory requirements. These factors are merged with capacity requirements planning (CRP), which delineates the production resources of the firm. Demand that exceeds production resources are used to determine potential needs for overtime scheduling, subcontracting, or back-order production runs. Long-term projections of continued demand that exceeds production resources may indicate the need for additional production facilities.

The aggregate sum of the above factors provide the information necessary to derive total demand for production. The MPS, while based on forecast data, removes uncertainty from production schedules by coordinating it with the master schedule. All supporting schedules are geared to the MPS.

Figure 8.1, which illustrates the manufacture of a product made up of an assembly and a component, is an example of the use of MRP. The demand for product A is established in the MPS. Subsequent and lower-level demands are derived and prioritized by MRP as a function of the MPS. This is illustrated by the sequential steps leading to the manufacture of product A. Items D and E are first combined to create assembly C; assembly C is then combined with component B to create product A. Component B is dependent upon product A and has a higher priority than items D and E, which make up assembly C.

Demand calculations by MRP require a bill of materials (structured parts list) that identifies the material that goes into a product. The *bill of materials* depicts the product at the highest or zero level. Assemblies that are a part of the product would be at level 1. Subassemblies, components, and parts would then occur at successively lower levels. The bill of materials reflects how the product is put together rather than simply listing the parts.

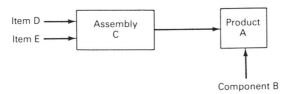

Fig. 8.1. An Illustration of MRP

Initially, applications of MRP considered only what was required to produce the product in developing order requests. For example, assume that 400 units of product A are to be manufactured and each product unit requires 1 unit of component B. This application of MRP considered gross requirements only. Therefore, 400 units of component B would be ordered without regard to what was in inventory or what may already be on order. Due dates and order placement times were determined by the lead time of component B.

The next refinement of MRP considered gross requirements in conjunction with inventory availability and what was on order and scheduled to arrive in time to support demand. The resulting net requirement was then placed on order. The determination of net requirements is illustrated in Table 8.1.

Table 8.1. NET REQUIREMENTS

Gross requirements		400
On hand	200	
Less safety stock	(50)	
Available in inventory		150
On order		50
Net requirements		200

This refinement of MRP resulted in very serious problems because an enormous amount of computation was required. This problem, however, was easily solved with the advent of computers. Computers solved this problem and led to the next step—time phasing.

Time-Phased MRP

A time-phased approach to MRP subdivides the demand into periodic production intervals. The illustration in Figure 8.2 takes the data presented in Table 8.1 and projects demand by weekly time periods. This product demand schedule is, in

Period (Week)		1	2	3	4	5	6	7	8
Projected Demand		50	30	60	0	70	50	80	60
Planned Receipts			50			10	50	80	60
On-Hand Inventory	150	100	120	60	60	0	0	0	0
Scheduled Orders				10	50	80	60		

Lot Size: Lot-for-Lot
Lead Time: Two Periods

Fig. 8.2. Time-Phased MRP

turn, supported by the identification of any scheduled receipts, a running total of available inventory, and the periods in which orders should be released. For example, prior to the start of production, 150 units are available in inventory. Product demand consumes 50 units during period 1, leaving 100 available in inventory. Period 2 demand is for 30 product units with a scheduled receipt of 50 units that were previously on order. Therefore, period 2 begins with 100 units in inventory, adds 50 units that come in, and consumes 30 via the production process. A net total of 120 units is available in inventory at the end of period 2.

At the end of period 4, 60 units are available in inventory. Period 5 demand is, however, for 70 units, which would lead to a negative inventory balance of 10 units. (The 10 units would actually be extracted from safety stock, which would protect against this type 1 uncertainty.) This projected negative inventory indicates that receipt of 10 units (lot-for-lot replenishment) must be scheduled to arrive during period 5. The lead time for replenishment is 2 periods, therefore, the order for these 10 units must be placed in period 3.

An important feature of MRP is its ability to react to changes in scheduling. The MPS reflects a total demand for 400 product units over periods 1 through 8, resulting in a production schedule encompassing 8 periods, as illustrated in Figure 8.2. A change in incoming receipts of material is compensated by changes in the production schedule within the constraints established by the MPS. For example, refer to Figure 8.3, which shows a revision of the product demand schedule presented in Figure 8.2. Figure 8.3 also assumes that the 50 units scheduled to arrive during period 2 have been delayed to period 4.

As before, production in period 1 consumes 50 units out of inventory, leaving 100 units available at the start of period 2. Period 2 production of 50 units leaves a total of 10 units in inventory at the start of period 3 (the anticipated receipt of 50 units has been delayed until period 4). Period 3 product demand is for 60 units; however, only 50 units are available in inventory. Disregarding the safety stock, the 10 product units scheduled for period 3 must be rescheduled for period 4 (or a later period) if the master schedule is to be maintained. The delayed 50 units arrive

Period (Week)		1	2	3	4	5	6	7	8
Projected Demand		50	50	60	10	50	50	80	60
Planned Receipts			50	→	50	10	50	80	60
On-Hand Inventory	150	100	50	(10)	40	0	0	0	0
Scheduled Orders				10	50	80	60		

Lot Size: Lot-for-Lot
Lead Time: Two Periods

Fig. 8.3. A Product Schedule Change

118 / *Material Requirements Planning* Ch. 8

during period 4, thereby permitting production of the rescheduled units, and MRP is again on track.

The preceding example illustrates that a lot size of one-for-one will ultimately reduce inventory to zero. At this time, scheduled orders are equal to projected demand and lag demand by an amount of time equal to lead time. Many materials, however, are either less expensive in quantity or are not provided in units of one. For example, a single nail would not be ordered in a one-for-one lot size. Figure 8.4 illustrates the previous example with a lot size of 100 units. The planned receipts in period 2 (50 units) may be attributed to an earlier shipment that was partially filled. This incoming material is the balance of that order.

Master Production Schedule

The *master production schedule* (MPS) is the primary input to MRP. It is a forecast of production in the same sense that a sales forecast represents a projection of future sales. They are related but are not the same. The firm begins with a sales forecast projecting the number of units it expects to sell or otherwise distribute. This sets the boundaries for the production function.

Production reviews the sales forecast to determine the quantity and profile of expected sales. (The sales profile reveals any seasonal or other types of variations that are expected during the forecast period.) This information provides manufacturing with the absolute quantity and the time intervals in which products are required if sales expectations are to be met. Manufacturing can now develop the production plan, which schedules production over time (to maintain reasonably level production rates), and can now review inventory, making adjustments as necessary. The master production schedule is then derived from production plans for each product included in the original forecast. This permits integration of the manufacturing activity with material and capacity requirements.

The master production schedule drives the production and inventory control system. It is the apex of the triangle, and lower-level conflicts are brought into focus at this point. The unexpected customer order that must be added to the schedule and the inventory that is carried to support the order that doesn't materi-

Period (Week)		1	2	3	4	5	6	7	8
Projected Demand		50	30	60	0	70	50	80	60
Planned Receipts			50			100		100	
On-Hand Inventory	150	100	120	60	60	90	40	60	0
Scheduled Orders				100		100			

Lot Size: 100 Units
Lead Time: Two Periods

Fig. 8.4. A Lot Size of 100 Units

alize result in conflict (and added expense). The risk of either occurrence is recognized and accepted by the firm as a concession to improved customer service. However, the master production schedule should not be changed to accommodate these or other perturbations if it is at all possible to continue with the original. The schedule that becomes totally unrealistic should, of course, be modified to reflect workable alternatives. This, however, does not obviate the fact that every effort should be expended to maintain the integrity of the master production schedule. To do less threatens the credibility of the entire system.

Capacity Requirements Planning

Master production schedules and time-phased MRP systems indicate what products are needed and when. This prioritization or priority planning must be followed by a determination of the capacity required to meet production objectives. Capacity requirements planning (CRP) begins with a tentative plan that identifies the needed capacity. The capacity is then compared with actual capacity to evaluate the potential of meeting the master production schedules.

Capacity requirements planning is very similar to the workload determinants discussed in Chapter 6. For example, 6 employees at 8 hours per day, 5 days per week have a theoretical capacity of performing 240 work hours each week. Thus, any combined workload of 240 hours or less lies within the capacity of this particular operation. A problem exists, however, in that capacity requirements are based on the time-phased MRP and are rarely of equal levels. For example, consider the product demand schedule presented in Figure 8.2. An assumed requirement of 5 hours per product unit would result in capacity requirements as pictured in Table 8.2.

Table 8.2. CAPACITY REQUIREMENTS FROM DATA IN FIGURE 8.2

Week	Product Demand	Hours
1	50	250
2	30	150
3	60	300
4	0	0
5	70	350
6	50	250
7	80	400
8	60	300

The average workload over the 8-week period is 250 hours per week, well within the capacity of the work force, assuming an average of 10 hours per week overtime. The actual time required per week, however, varies from 0 to a maximum of 400 hours. No facility can operate with this type of variance. The demand per week may have to be varied within the constraints of the MPS to provide a more realistic level of activity. More reasonable variations than those presented in this example could, perhaps, be compensated through limited use of overtime, a review of alternate tasks that are the responsibility of this work force, or other measures appropriate to the firm.

Elements of Success

MRP is a rigorous management discipline that demands an overall commitment if it is to succeed. It must be approached with a great deal of education and exposure. MRP has revolutionized production and inventory control concepts, yet, without the dedication of those who use the system, it is no better, and in fact may be worse, than its predecessors.

Success within the logistics arena is difficult to quantify, and production and inventory control is no exception. MRP success is, however, somewhat easier to measure since improvements may be noted in various areas, such as in

A. Stockroom accuracy (often exceeding 50 percent)
B. Reductions in the quantity of unplanned material issues (up to 70 percent and higher)
C. Reductions in the number of "short loads" received on the shop floor
D. Productivity as evidenced by frequently meeting or exceeding production goals

These improvements and others offer dramatic testimony to the potential inherent within MRP.

Beyond Material Requirements Planning

Material requirements planning led to tremendous advances in production and inventory control technology. Inventories decreased in size and complexity, and inventory managers were aware of both the quantities and locations of stock entrusted to their care. This knowledge of what was available, coupled with realistic estimates of the quantities needed to meet production goals and with the time phasing of that need, enabled a disciplined system, which, for the most part, assured the timely acquisition of needed additional materials.

Yet there was still an inventory. Inventory stock still turned up missing or reappeared unexpectedly when no longer needed. Promised delivery dates were still delayed, occasionally to the point where changes could no longer salvage the production schedule, and surplus work in process would sometimes accumulate on the shop floor. MRP is a gigantic step forward, but it isn't the panacea it is so loudly proclaimed to be by some of its more avid proponents.

MRP, with all its refinements, remains a traditional production methodology. Incoming parts and material are still unloaded from transport vehicles and put in storage areas. Subsequent demand, as derived from the production schedule, leads to the right part or quantity of material being sent to the assembly line. Here, the parts and material are used in the creation of subassemblies, which, in turn, progress down the line. Eventually the finished product is ready for transfer to the marketing function.

A superior system would eliminate inventory by having each item that is required during production appear when needed and in the exact quantity needed. Such a system would require several changes in current concepts of doing business. Two of the more drastic areas of change would be the following:

A. Quality control of the parts and materials used in the production process. The idea that some small percentage reject rate is acceptable must be dispelled. A system that delivers the right item to the right place at the right time and in the exact quantity needed demands a zero defect philosophy and practice.
B. The relationship between the firm and its suppliers. The firm must have an assured source of the parts and materials required during the production process. Credibility and confidence within the firm must be based upon on-time deliveries of items that exhibit the right quality level. The supplier must become an extension of the assembly line. The supplier, on the other hand, must have the confidence of continued support from the firm.

This superior system does exist. It is referred to by various titles such as *Stockless Production, Zero Inventories,* or *Just-in-Time,* and its fundamental objective is the elimination of waste from the manufacturing process.

Just-in-Time

The just-in-time (JIT) concept is not new. Rather, it is an old concept with a renewed focus on having the right quantity of the right items of the desired quality level at the right place at the right time. As may be imagined, JIT demands a great amount of discipline and control if it is to be implemented successfully. One of the more desirable features of JIT is the elimination of inventories and the associated inventory, storage, and control costs.

The American interest in JIT began to emerge when consumers initiated questions asking why higher quality Japanese products were selling for less than similar products made in the United States. The initial premise that lower labor costs were responsible lost a lot of credibility as U.S. industry relocated factories overseas to avail themselves of this same "cheap" labor. However, "cheap" labor was not reflected in the cost of goods, and the quality level was still perceived as less than optimum.

Wage rate arguments received another blow during the oil crisis of the 1970s. A tremendous increase in transportation costs more than offset any wage advantage, yet Japanese products were offered at acceptable prices and at a quality level perceived by many as being superior. As more and more Japanese products were consumed, people continued to question their superiority in quality and costs.

The answer was found in the Japanese mastery of American manufacturing concepts. Concepts largely rejected by American manufacturers allowed the Japanese to produce massive quantities of quality products and market them far beyond their own shores. This success was due to an application of the JIT philosophy.

A Contrast of Philosophies

United States industry entered the first half of the twentieth century as a creative giant and world leader in technology and innovation. The latter half of the century, however, witnessed the beginnings of a downturn as this awesome force began to weaken. During this same period of time a market-oriented economy began to emerge, and the consumer demanded quality with no particular regard concerning where the product was produced.

Consumer dissatisfaction with American products was most evident in the automobile industry during the late 1970s and early 1980s. An entire industry was literally brought to its knees by a combination of the economy and an acknowledged preference for Japanese and European cars. This sent a strong signal throughout industry that it was time to reevaluate its practices or become resigned to being a loser in the marketplace. The difference in manufacturing philosophies came under intense study as the need for change was now a necessity.

The most visible difference between the two philosophies is the method in which inventory is acquired and used in the manufacturing process. The traditional method gets behind the material and pushes it through the manufacturing process until it has been transformed into the finished product. Under the push system, material is manufactured into subassemblies and either pushed up to the next level of assembly (whether needed at that time or not) or moved into a stockroom until needed. A given work center may even remove the stored assembly for additional work, then return it to the stockroom once more. The reasons behind this kind of system include

A. Security and protection against damage
B. Flexibility against unforeseen variations in demand (type 1 uncertainty)
C. Protection against rejection or failure
D. Protection against delays in material deliveries (type 2 uncertainty)

The Japanese, instead, rely on a demand-pull system, whereby products are removed one at a time as they are completed. This, in a sense, creates an opening within the manufacturing process that "pulls" the exact amount of materials into production. The Japanese demand-pull concept is driven by a card-based system called *Kanban*. This assures that materials can only be provided when needed as evidenced by a request from another work center.

Another difference in philosophy can be traced to the American characteristic of running the machines of production as fast as practical during work hours. Any excess production is sent up the line as shop floor inventory or sent to storage as added inventory. Excess inventory is justified as necessary to maintain the manufacturing flow should the equipment of production break down, thereby interrupting the production process.

The Japanese, on the other hand, have concluded that interruptions of the production process, such as those engendered by a regular program of preventive maintenance, are beneficial. Preventive maintenance interruptions (cleaning, adjusting, oiling, inspecting, and so forth) lead to equipment that produces higher quality items more consistently and over a longer period of time than equipment operated under the American philosophy. As a result of this philosophy, the amount of unpredictable out-of-service time encountered in Japanese factories is significantly less than that experienced in the United States.

Implementing JIT

Traditional U.S. manufacturing practices views production and material ordering by the lot as very desirable. This is in sharp contrast to the Japanese method, which has achieved great success by producing in smaller quantities and by ordering materials more often and in smaller quantities. The concept of not producing items or buying materials until they are needed must be adopted as a prelude to implementing any program such as JIT.

Another attitude of U.S. manufacturers that may even be more difficult to alter involves the very natural resistance to change. Manufacturing lines prefer long, uninterrupted runs of the same product. This is due, in part, to the desire for increased productivity and efficiency of operation. Another factor, however, is the comfort derived from the routine and familiar.

JIT views production runs from a totally different perspective. Smaller product runs mean that changes in the production line are performed much more frequently. The objective is not to reduce the number of changes. Instead, it is to reduce the time required to make the change. A goal of ten minutes per work center is considered reasonable. This, however, must be supported by the design and layout of the work center. For example, any tools or attachments that are required for machine setup procedures should be convenient and accessible. Guidemarks on machinery, setup diagrams, and color coding are other factors that may be used to accelerate the change to alternate product runs.

A regular program of preventive maintenance is the next issue confronting manufacturing management. The derived benefits of a decrease in unscheduled (and unpredictable) downtime, longer machine life, and a more consistent output of a higher quality level more than offset the cost of a preventive maintenance program.

Layout of the production facility is an issue that also must eventually be considered. The manufacturing floor, as viewed from above, should present a linear and efficient flow of work centers. Manufacturing activities should follow this flow as material, subassemblies, and assemblies move through the production process. Color coding of work centers make them easily identifiable, thereby facilitating this flow. JIT emphasizes the single-unit nature of this center-to-center flow (one product or a small container of products), which takes place only at the request of the receiving work center.

Equipment flexibility and demand-pull contribute to a system that produces only at the rate of consumption. Manufacturing cycle times require adjustment only to assure that successive work centers are in synchronization (no work center has to either wait for or store an item of production). Cycle time changes, when necessary, are implemented through activities such as

A. Variations in the speed of production equipment
B. Changes in the work force at individual work centers, requiring a work force that has the ability to perform in a variety of manufacturing activities
C. Changes in machine setup times
D. Reevaluations of manufacturing lead times

This definition of work centers leads to capacity planning that is, in fact, labor dependent. The impact of overcrowded inefficient work space is not an issue.

Quality Function

Quality, which implies that an item is reliable and fully capable of functioning as intended, is a critical factor in the JIT environment. Quality must begin with the materials of production; therefore, suppliers to the firm take on a greater significance. The supplier must become a part of the "manufacturing family" with an interest that extends through to the finished product. This approach typically leads to a smaller number of suppliers, with each receiving a larger share of business from the firm.

As may be imagined, the selection of suppliers is extremely important to the firm. Preliminary selections may entail buyer certification and a trial period during which both quality of the product and reliability of the supply are evaluated. JIT may even lead to an inducement for suppliers to relocate near the firm. The nature of the supplier-firm relationship under JIT certainly offers an advantage to those in proximity to the firm.

Quality in the manufacturing environment is an equally important factor. JIT, in order to assure a quality product, inspects the first and last item produced by each work center. Any variance from predetermined standards is identified on work center control charts. Inferior quality detected during the production process may lead to a shutdown of the entire line pending identification of the cause.

A quality-oriented attitude is maintained throughout the work force by a combination of feedback and information concerning the role of each worker in relation to the final product. Each worker is aware of part and product reliability, what failed, why it failed, and the costs associated with repair. This sharing of information intensifies worker interest in product quality.

Product design techniques represent another area that must change in the JIT environment. Products must be designed with mass production in mind. Design engineers must consult with manufacturing engineers throughout the design process. Logistics and logistic-related disciplines such as reliability and maintainability engineering must be included to enhance customer satisfaction following the transfer of ownership. All manufacturing and manufacturing-related plans and procedures must be completed prior to the start of production. Common and high-quality materials and parts should be used throughout the production process.

Initial production runs of the new product should include design engineering to enable on-the-spot engineering changes as necessary. This time should also be characterized by detailed records of manufacturing factors such as cycle times, stock movement, problems that are encountered, and the solutions to those problems.

What Is Kanban?

Kanban, which means "card," is a method of controlling stock flow and the activities of each work center. Kanban requires manufacturing equipment reliability and well-defined linear flow of the production process. Additionally, each work center must have an input stock point and an output stock point.

Kanban utilizes two types of cards: a move card and a production card. *Move cards* authorize the movement of one container of items (or a single item) from the output stock point of the producing work center to the input stock point of the requesting work center. The card includes information such as the item part number, container quantity, the card number, the producing work center, and the requesting work center. Production cards authorize the production of the items needed to resupply the work center's output stock point. Information on the production card indicates the item to be produced, container capacity, the part number, the card number, the bill of materials, and the supplying work center.

Production cards are placed in boxes located in each work center at the beginning of each work day. Each worker gets a card that authorizes the production one container for storing at the output stock point. A work center requesting completed items acquires them by use of a move card. The production card is then returned to the supplying work center's production card box, thereby authorizing the production of another container of items.

The requesting work center places the move card in the work center move card box when the first item is removed from the container. An empty container results in the use of this move card to obtain another container.

The Kanban system controls the inventory level of any part at any location on the manufacturing floor and allows the worker to produce only upon request. This requires a high degree of coordination in that production cycles must permit the request and utilization of a container to occur in the same amount of time as that required to produce it.

SUMMARY

Production and inventory control is a disciplined and integrated system that focuses on four fundamental issues:

A. Planning for priority
B. Planning for capacity
C. Control of capacity
D. Control of priority

Priority planning for the make-to-order-one-piece firm is relatively simple because due dates are generated by customer orders. The make-to-stock may use the time-phased order point; however, with dependent demand components, material requirements planning (MRP) is the preferred choice. Neither system, by itself, provides assurances that derived priority plans are valid. A master production schedule (MPS) must be used to keep priorities in line and to maintain the credibility of the formal system.

Capacity planning is a means of evaluating the magnitude of projected activities, to determine, by time phased periods, capacity required versus that available. Both differences between the two and undue variations of manufacturing periods must be resolved prior to the start of production.

Production and inventory control, while representing a giant leap forward in manufacturing technology, does not represent the final solution to production problems. The next evolutionary improvement is to be found in the just-in-time (JIT) system.

The JIT philosophy may be expressed as the process of having the right quality item at the right place at the right time and in the exact quantity needed. Production, under JIT, takes place only upon request and only for the amount that is needed.

JIT is dedicated to the elimination of waste in the manufacturing process and has the potential of revitalizing the firm that chooses to adopt it. The JIT objective can be translated directly into improved product quality, improved reliability and maintainability, and increased consumer satisfaction.

QUESTIONS FOR REVIEW

1. Cite an example of an informal system that operates in industry.
2. Provide a brief description of priority planning.
3. How does material requirements planning differ from order processing?
4. Explain the difference between a sales forecast and a master schedule.
5. What is capacity planning?

6. Is safety stock of less importance under an MRP system? Explain your answer.
7. What are hot lists and expediting?
8. What are the objectives of the just-in-time system?
9. What is the function of a move card?
10. How does JIT eliminate inventory?

9

THE ERA OF CUSTOMER SERVICE

Integrated logistic support and the ILS manager who coordinates and orchestrates the variety of elements that make up this discipline, offer performance as a product. Performance and the services inherent therein are summarized as follows:

A. The availability of the needed (desired) logistic elements
B. The capability residing within those elements
C. The quality of the service that is provided through logistics performance

Virtually any level of logistics performance (service) is possible if the firm is willing to assume the associated cost. The service to be provided by logistics is, however, bounded by the firm. It falls to the ILS manager to achieve the balance between performance and cost that best meets the goals established by the firm.

Logistics as a Service

Logistics is a service function. It exists to provide a service to the firm or to customers of the firm. Service, to either the firm or to its customers, is provided through the application of the various elements, which, in toto, are grouped under the logistics umbrella. Logistic service within the firm emphasizes the elements of material management, internal inventory, and physical distribution. It is not, however, restricted in any way to this narrow range of services. To the contrary, the very nature of these primary functions dictate varying degrees of emphasis on the other elements of logistics. For example,

A. Each of the three primary functions—material management, internal inventory transfer, and physical distribution—directly lead to the requirement for an inventory. The materials of production are acquired in excess of need, and this excess must be controlled and accounted for. Work in process, whether within a single firm or within geographically dispersed operations, must be controlled, accounted for, and processed through the mechanism of internal inventory transfer. The products of production are, except for the rare make-to-order firm, produced in excess of demand with the surplus being stored as a finished goods inventory. An inventory is also required for storage and control of the spare and repair parts that are required in maintaining production equipment.
B. The presence of inventories necessitates the establishment of facilities. Warehouses and stockrooms must be provided for storage and control of the material for production, the products of production, and other inventory items that are required by the firm.
C. Production equipment will fail, and, upon failing, it must be restored to operational service in the shortest possible time. This necessitates the establishment of a maintenance and repair capability and the requisite facilities for maintenance.
D. The manufacturing work force must be trained in the operation of production equipment. Maintenance personnel must be trained to develop a repair capability for the same equipment. Operation and maintenance responsibilities, in turn, require the use of technical publications, tools, and test and support equipment.

From the preceding description it can be seen that the firm, in the normal conduct of business, requires the support provided by each logistic element. It is true that a given logistic element—maintenance and repair of the equipment of production, for example—may be obtained from an intermediary specialist rather than within the firm itself. This, however, does not obviate the need for this service; it is simply obtained from an external source. The responsibility for the quality of this support is, however, a responsibility of logistics within the firm.

Logistics plays an equally significant role in providing support to customers of the firm. This is product support following the transfer of ownership, and it has a slightly different emphasis when compared with support within the firm. Maintenance and repair of a product are functions of customer demand.

Logistics concerned with this demand generally extends beyond the boundaries of the firm. Additionally, complex products may require customer training in product operation. Limited training, as in the personal computer market, for example, may be included in the acquisition cost. Normally, however, additional training is offered to the customer as a product in its own right.

Logistics product support extends to operating manuals, which are provided with virtually every product. Frequently, maintenance publications are also offered, usually at an additional cost. Logistics in support of the product does not exclude those elements that are emphasized within the firm. For example, the maintenance and repair activity requires a facility and an inventory to meet its objectives. An additional inventory may be set up as a spare and repair parts source for the consumer. These inventories and facilities require the same degree and type of control as similar facilities within the firm. Physical distribution and related transportation needs also apply to product support as, for example, in the transfer of defective items to and from the repair facility.

Logistics, and the support it provides, is a critical resource of the firm. It must be realized, however, that this support has the ultimate objective of reducing costs and/or enhancing profit of the firm. Customer service is expensive, but the payoff can be high in both repeat and new business. The chances are very high that unhappy customers will not express their displeasure directly to the firm. The chances are equally high that they will not return to the firm and will tell from ten to twenty others of their dissatisfaction. A moment's reflection on the pyramiding effects of this quickly reveals the value of customer service.

What and Who Is the Customer?

The firm exists within its environment. It draws resources from that environment (materials of production and the work force that manipulates those materials) and sells the products to customers residing within the environment. From this it is but a short step to the logical conclusion that customers are the reason, the only reason, for the firm's existence. The alternative, a firm producing goods for the sake of production, is too ludicrous to contemplate. A customer, then,

A. Is always the most important person in the firm and must be accorded treatment that befits that position, whether in person, by phone, or by mail.
B. Is not an interruption of work but the reason for it. The firm is not doing the customer a favor by offering service. Rather, the customer does a favor to the firm by offering the firm the opportunity of providing the service.
C. Is the individual who brings needs and desires to the firm. It is the responsibility of the firm to handle them so that the firm and the customer profit.

The customer, however, is more than the individual who purchases the products of production. He or she must share the distinction with others in the environment and in the firm.

The External Customer

The ultimate customer is the end user or consumer who purchases the product for his or her own use. This customer represents the end of a long and multifaceted logistic channel that begins with the suppliers to the firm. The supplier, who in fact may be the customer of still another supplier, initiates the logistic channel by providing the materials of production to the firm. The firm, in turn, produces the finished good, which is stored in a warehouse pending demand for the product. The warehouse provides the item to a wholesale merchant, who transfers it to the retailer. The retailer, representing the end of the logistic channel, satisfies consumer demand by offering the product for sale. Thus, the consumer is the final authority and his or her decisions in the marketplace determine life or death of the firm.

The consumer began this rise to power in the years following World War II. The increased availability of consumer goods and competition, which provided the alternative of a choice among similar products, endowed the consumer with decision power over the firm. The firm that refused to acknowledge and accept the needs and desires of the consuming public could no longer grow and prosper in a free market society. Industry, like it or not, was in an era of market orientation.

Rigorous competition, the consumer's ability to select from alternative choices, and the very natural tendency of imitating more popular products led to a high degree of product similarity. Differences were largely artificial, created by cosmetic enhancements or the advertising industry.

These circumstances led directly to the era of customer service. The products may be similar or even identical; however, there could be major differences in the attitude accorded customers and in the type of product support that was offered

---

Product Support and Guarantee

Product support is more than just words in a magazine ad or a brochure.

True support involves time and energy, and a genuine concern for customer satisfaction. Many software companies claim to offer it, but few really do.

We've dedicated ourselves to offering the absolute best product support and guarantee in the industry.

Our 24-hour Hotline is really answered 24 hours a day, and our "NO FINE PRINT" Lifetime Guarantee does in fact last a lifetime.

And best of all, both are INCLUDED IN THE PRICE OF THE SOFTWARE, not offered as an add-on or sold on a year-by-year contract basis.

We also offer free updates if a product is revised or changed. All you have to do is ask, and send us your master program diskette.

In our view, if YOU have a problem, WE have a problem. The Friendly Ware® "No Fine Print" Lifetime Guarantee states simply:

If our master diskette fails due to normal use, FriendlySoft, Inc, will replace it AT NO COST within 48 hours or receipt of the original master diskette.

If our master diskette fails due to abnormal use (food on the hub ring), we will replace the programming onto your diskette AT NO CHARGE within 48 hours of receipt of the original master and a blank, unformatted diskette.

Reprinted courtesy of FriendlySoft, Inc, Arlington, Texas (used with permission).

following the transfer of ownership. The previous article, reprinted from the FriendlySoft, Inc., product brochure and used with permission, is an illustration of this type of product enhancement.

Does the degree of support described in FriendlySoft's guarantee result in a sufficient quantity of added business to justify its cost? It is very difficult to measure. With this approach the firm has a defined level of business. What would the level of business be if this support was not provided? This too is a difficult question to answer since the support is there. Any answer must be largely subjective, since there are many other factors that impact consumer choices. In general, all other things being equal, the extra customer support will yield increased sales over time.

The Internal Customer

A customer is much more than an end user of a product. The end user is, in fact, a relatively rare type of customer to the majority of firms, although work is performed in response to customer needs. The retail merchant is usually the only firm having the consumer as a customer. The retailer is a customer of the wholesaler who is a customer of the firm. The firm, in turn, represents the customer to a host of suppliers. Similarly, the U.S. government is a customer of the firm that produces large systems for government consumption.

The customer of the firm is not, however, the customer of the workers in the firm. The customer of the worker is the individual or group for whom the worker is performing an activity. For example, the customer of the manufacturing employee toiling within a work center of the JIT system is the employee at the next work station. Customers of the ILS manager include the various functions and groups that need and desire the services of one or more of a logistic elements.

Logistics, as a service function, has as its customers each function within the environment of the firm that is seeking the support of a logistic activity: the marketing manager who requires support of product distribution, the manufacturing manager who needs inbound transportation for the materials of production, the function requiring an inventory. All are customers of logistics. These functions of the firm, or individuals within the firm, are customers in the same sense that the consumer who purchases an item from the retail merchant is a customer. And, as customers, they provide the demand for each of the various logistic activities.

A Service Policy

Given that the firm is going to provide service to its customers, management intervention is required in determining the degree of that service. What approach is the firm going to take toward its customers? What level of service is to be provided? As previously indicated, virtually any degree of service is possible if the firm is willing to underwrite the cost.

Implementing the logistic element implies some level of performance (activity) that begins with a request for service and culminates with the satisfactory completion of that request. This approach to logistics performance is illustrated in Figures 9.1 and 9.2. Figure 9.1 represents logistics within the firm. First a need is identified and defined (for example, defective equipment is in need of a part). That need is then transmitted to the appropriate logistics function (the employee walks to the stockroom and requests the needed part). The request is then processed, fulfilled, and delivered to the customer (the inventory processes the request, obtains the part from inventory, and hands it to the requesting employee). The need has now been satisfied, and the logistics performance cycle is complete.

Figure 9.2 illustrates this same concept as it applies to resources external to the firm. In this example, material management identifies a need for material and transmits that need to a supplier. The supplier receives the order for material, processes and fulfills the order, and delivers the material to the firm. This example would also serve to address the wholesale merchant who requests a shipment of finished goods as replenishment inventory. Requests for other logistic services follow a similar performance cycle.

Note that both examples included identical cycles of logistics performance. In either example, each cycle incorporated the factors of recognition, identification, processing, fulfillment, and delivery. The desired level of this logistics performance is defined by management policy. It then becomes a function of logistic capability, availability, and quality.

Capability

Capability is measured by a combination of the relative speed and consistency of the logistics performance cycles illustrated in Figures 9.1 and 9.2. Given two or more suppliers, the one with the shortest performance cycle time and the most consistent performance cycle duration would be considered the most capable. Either of these two attributes may be considered to be the more important to a given firm. For the firm in our example, speed of performance is the primary requisite. Thus a performance cycle that is faster but less consistent is considered more capable. Capability must be evaluated within the context of both the needs of the firm and the appropriate methodology for meeting those needs.

Availability

Availability is the capability of the logistic system in providing service on a predictable basis. This definition specifically excludes the establishment of a discrete time interval in evaluating availability. (Time is the province of logistic capability.) With the exception of certain perishable goods or the stated desire of customers, speed of delivery (or the lack thereof) is not an availability factor. The important availability issue is the predictability or consistency of service or, stated another way, the probability that a needed item or service will be available upon request. For example, the manufacturing manager needs to know when the material of

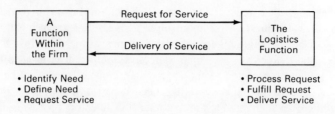

Fig. 9.1. Logistics Performance Within the Firm

production will arrive with a high degree of certainty. Whether it arrives in two weeks or in six weeks is much less important than the stability of the time between arrivals. Extended periods of time will, of course, lead to an increase in the average level of inventory within the firm.

Availability within the logistic system is a direct result of inventory stock and the established safety stock policy. It may be measured by determining the percentage of requested items that are out of stock relative to total inventory stock. As an illustration, assume that 100 individuals request an item from inventory. Of these 100 individuals, only 98 receive the requested item and 2 do not because the requested item is out of stock. Two percent of the items are out of stock, therefore the inventory availability is 98 percent.

There is a problem with this measurement technique, however, since it does not consider the velocity or turnover of individual items. A rarely requested item may be out of stock for extended periods of time without presenting a problem. A stock-out of a fast-moving item, on the other hand, may lead to problems within minutes. For this reason, preferred measures of inventory availability incorporate an analysis of performance over time.

Quality

Performance *quality* is a function of the quantity of damaged, missing, or incorrect items provided by the logistic service. Measures of quality, as of capability and availability, must be determined over time. It would be grossly unfair to assign a quality rating to the logistic practitioner on the basis of a single performance cycle or on a very limited sample.

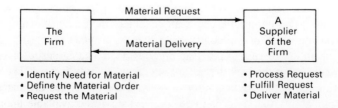

Fig. 9.2. An Example of Logistics Performance Extending Beyond the Firm

Planning a Customer Service Policy

The goal of integrated logistic support is to design a package of logistic resources that provides a level of performance capable of optimizing the goals of the firm, such as

- A. Maximizing profit. This is the most common goal of the commercial firm.
- B. Minimizing losses. This is (hopefully) a short-term objective undertaken in an attempt to restore profitability. Losses are minimized through a reduction in variable and, where possible, fixed costs.
- C. Maximizing service. This is rarely if ever, an objective of the firm. Providing the maximum degree of service possible under specific constraints levied by management is, however, a relatively common goal.
- D. Gaining a competitive advantage. This is a short-term objective that is occasionally implemented under special conditions. The prosperous firm, for example, may "dump" its products on the market at prices below cost in order to attract the customers of competing firms. Obviously this practice of selling at a loss cannot be continued indefinitely.

For the most common objective—the maximization of profit—the solution is extremely simple in theory. All that is necessary is to continue to raise the level of customer service (through increased expenditures) until marginal cost is equal to marginal revenue.[1] This is, unfortunately, virtually impossible to accomplish in practice. Significant factors that inhibit a determination of marginal equality include (1) the difficulty of accurately measuring incremental cost within the logistic element, (2) the practical impossibility of adding one more unit of output to the logistic activity, and (3) the inability of determining the impact of logistics performance on profit. The following examples illustrate the difficulties inherent in accurately measuring the contribution of logistics.

- A. The rarely requested item in inventory has a cost that can be determined. What is the cost in lost sales if it is requested and not available? What is the increase in profit that would have accrued if it had been available?
- B. Under the assumption that one unit of the material of production is equal to 100 pounds, what is the contribution to profit that results from the addition of one additional unit?
- C. There is a stock-out in the material of production that extends over one eight-hour workday. Product inventory includes a supply of the finished good equal to five days of normal demand. What is the deduction from profit that results from this stock-out?
- D. A training course is added to assist the consumer in operating the product. What is the impact on profit that can be attributed to the class? What is the

---

[1] *Marginal cost* is the extra or additional cost of producing one more unit of output. It can be determined by noting the change in total cost that is created by production of that unit. *Marginal revenue* is the increase in revenue that is provided by the sale of one more unit.

difference between current profits and the profit that would have been realized had the class not been offered?

From the preceding, it is apparent that it is very difficult to assign an accurate dollar figure to the impact of logistics on profit. This difficulty is compounded when a change in the level of logistics performance is considered. For these reasons, the setting of customer service levels is largely a subjective decision. There is, however, a degree of objectivity that can be injected into the decision process, thereby providing some data on the monetary results of the decision. The objective elements permit an approximation of the ideal logistic system through (1) a determination of the least-total-cost design and the level of logistics performance associated with that design and (2) sensitivity analysis of changes in service levels in terms of cost and revenue contribution.

Least-Total-Cost Design

Least-total-cost design begins with determining the combination of logistic resources that has the lowest total cost (both fixed and variable). This determination is a pure cost trade-off and does not consider levels of logistics performance. Instead, the level of performance possible at the lowest total cost establishes the reference or threshold service level, the minimum level of service that may be expected of the logistic system. For example, the lowest cost approach to physical distribution considers minimum transportation charges only. Factors such as the speed and consistency of service (capability) and the predictability of service (availability) are not included in the analysis. Capability and availability are only as effective as the minimum transportation costs allow.

In a noncompetitive environment, the level of service afforded by the lowest cost approach may be all that is desired, and, in fact, an approximation of this level was not uncommon during the era of the production-oriented firm. This "you will have it when you get it" attitude is obviously impractical in a market-oriented society. The customer would simply procure the item elsewhere.

Since the level of logistics performance described above is unrealistic, the next step is to determine what the acceptable level of performance should be. This determination is a function of the degree of risk that management is willing to assume. For example, assume that a decision is made to establish an inventory with an acceptable stock-out level of 90 percent. This means that, on the average, 90 percent of the people who request an item from inventory receive the item. Conversely, this also means that 10 percent of the people will not receive the item. To evaluate the effects of this decision, assume the following cost elements:

A. The average cost of all items in inventory is $100 per item.
B. Inventory carrying costs add $25 to the cost per item (including personnel staffing costs).
C. The average sale price of an inventory item is $175.

Lost sales due to stock-outs at a 90 percent inventory level are therefore equal to

175×10 or $1,750 for each 100 customers. Deducting the total inventory cost of $125 per item, the resulting loss in profit is $1,750 - $1,250 = $500. This seems to indicate the desirability of increasing the level of inventory (establishing a stock-out level at some figure greater than 90 percent).

However, the decision is unfortunately more difficult since the probability of the increase in profit must be taken into consideration. To illustrate, the increased profit was generated by the sale of 10 additional units at a price of $175 per unit. The probability of this is, however, 10 percent, resulting in expected revenue and profit of $175 and $50 respectively. This gain must be measured against the additional cost of 10 more items in inventory ($1,250). Does opportunity for increased revenue justify the additional cost engendered by an increased stock-out level?

The cold logic and strict mathematics of this example excludes the potential of lost sales and dissatisfied customers impacting future sales. Subjective decisions of these "costs" by management must now be entered into the equation to evaluate the merit of an increase in the level of customer service.

Sensitivity Analysis

Sensitivity analysis is a determination of the difference in cost between two possible alternatives. Sensitivity analysis within the logistic system begins with the reference (threshold) level of service established by minimum cost analysis. The level of service is then varied from that reference by changing (1) the number of logistic facilities, (2) one or more factors of the logistics performance cycle, and (3) safety stock levels.

The number of facilities within the logistic system determines the level of service attainable if the performance cycle and safety stock are held constant. Take, for example, a warehouse operating 24 hours a day with a single delivery vehicle and having a 6-hour performance cycle to a market area comprised of 20 retail outlets. This warehouse could, on the average, provide 24-hour service to no more than 20 percent of the market: One delivery vehicle and a 6-hour performance cycle permit deliveries to only 4 of the 20 retail outlets in a 24-hour period. A maximum of 40 percent of the market would receive 49-hour service, and 60 percent would receive 72-hour service.

The addition of another warehouse and delivery vehicle (or the addition of another delivery vehicle to the existing warehouse) would increase the level of service to 24 hours for 40 percent of the market. This relationship is illustrated in Table 9.1. Note that there is no requirement to exceed five facilities since this number achieves maximum performance within performance cycle constraints. Note that the incremental cost associated with each addition to the logistic system increases at a linear rate (assuming the marginal cost of each additional facility is the same). This dramatic increase in cost must be carefully considered in relation to the increase in service that is gained by each addition.

The speed and consistency of performance cycles are also factors that may be

Table 9.1 LEVEL OF SERVICE AS A FUNCTION OF THE NUMBER OF FACILITIES

Number of Facilities	Service Level		
	24-hour	48-hour	72-hour
1	20%	40%	60%
2	40	80	100
3	60	100	100
4	80	100	100
5	100	100	100

modified when evaluating changes in the level of logistics performance. For example, air freight may be used to decrease performance cycle time, thereby improving customer service. This approach to increasing customer service (increased performance cycle speed) is characterized by significantly higher variable costs, lower average inventory levels, and greater system flexibility. By way of contrast, the addition of logistic facilities, such as an additional warehouse involves a higher fixed cost and a relative decrease in system flexibility.

A third alternative is to vary the level of customer service through changes in the amount of safety stock. Increases in safety stock levels decrease the possibility of stock-outs. The advantages of this increase in availability must, however, be weighed against the cost of the additional inventory, which can be substantial.

Implementing a Customer Service Policy

Implementing a specific level of customer service is a major management decision concerning which many managers are unduly optimistic. An increase in availability of a few percentage points or a decrease in the performance cycle time of a few hours may represent an enormous cost impact. This is particularly true if service is already being provided at a relatively high level. For example, consider a firm engaged in product distribution that provides 48-hour service to 92 percent of its customers. The firm decides to improve service through an increase in the number of warehouses within the logistic network so that 95 percent of all customers receive 48-hour service. This 3 percent improvement in customer service may require five new warehouses at a cost of $2.0 million each with a total cost of $10 million.

The additional storage facilities reduce the average inventory at each location (a cost reduction). Inventory costs of the total logistic system, however, increase, since an inventory is now required for five additional facilities. Similarly, the average transportation cost per facility decreases, since more facilities are serving the same market area and the average distance per shipment has decreased. (This discussion, of course, ignores the effects of new business generated by the improvement in customer service.) The net result of these cost variables may act to reduce the effective cost of new facilities by $1 million. This leaves a total cost of $9 million, which must be offset by new business.

More information regarding the firm must now be given to properly evaluate the decision to improve its customer service:

A. The firm is firmly entrenched within the market, with a 28 percent market share.
B. The firm realized a before-tax profit of $43 million last year on sales of $391 million.

The return on sales represents a before-tax profit margin of approximately 11 percent. The additional five warehouses must generate approximately $82,000,000 in new sales to justify the improvement in service on a break-even basis. This represents an increase in market penetration of almost 6 percent. Is this likely? Management must arrive at a decision to accept or reject any change in customer service policy as a function of potential markets and the degree of risk the firm is willing to assume.

Previous changes in customer service consisted of modifications of the structure of the logistic system: Warehouses were added or removed, the transport vehicle was changed to vary performance cycle speed, and safety stock was varied above and below a previously established policy level. Each variation led to a change in the level of customer service offered by the firm. There are, however, alternate methods that could have been utilized in changing the level of customer service. These methods involve the operating procedures of the firm rather than the logistic structure.

Alternatives to Changing the Logistic Structure

Previously cited changes to the structure of the logistic system are an effective method of improving the level of customer service. They do, however, require major financial resource commitments and rely upon a substantial increase in sales for economic justification. Increased sales must come from new customers who are attracted to products of the firm or from increased market penetration which attracts customers from entrenched and capable competitors. Either approach entails the expenditure of large sums of money for promotion and advertising, which effectively increases the cost of structural changes.

Management may wish to improve the existing level of customer service, yet it may hesitate to commit the resources of the firm on the basis of market projections. Fortunately, there are alternatives that involve considerably less expense and that may be as effective as the new warehouse in improving customer service. Many of these options are little more than good business practice and may actually reduce cost while increasing the customer base through improved service.

Consider this dramatic example of one approach to an economical improvement in customer service. A new car dealer could have various models and product options available, constituting a potential inventory of thousands of product units. This vast inventory far exceeds the capabilities of even the largest retail dealer. Individual dealers, therefore, maintain a very limited inventory of the products (new cars) that are available for sale to the consumer. Given the limitations of this inventory, it is likely a prospective buyer could desire a model with selected options that is not available. He or she may be equally averse to waiting six to eight weeks for delivery from the factory, and the sale may be lost to a competing dealer or even to another manufacturer.

The car buyer is not, however, limited to the selection available in the dealer's facility. A simple communications link ties dealers within the state and within a multistate region into a vast automobile inventory resource. The new car, unavailable at the local dealer, may be found within this inventory at-large and may be available for delivery to the customer in a matter of days (instead of weeks as when ordered from the manufacturer).

The added flexibility created by this simple communications network provides the dealer with an effective increase in inventory, thereby increasing customer service.

Another example of improved customer service is illustrated in Figure 9.3. In this figure, the firm supplies a market area consisting of a major customer who orders in truckload (TL) lots and several customers who purchase in smaller quantities. The smaller orders are delivered from a warehouse using economical local transportation rates. The warehouse purchases TL lots from the firm, attaining economical transportation charges. The major customer, who buys in TL lots, bypasses the warehouse and receives products directly from the factory. The

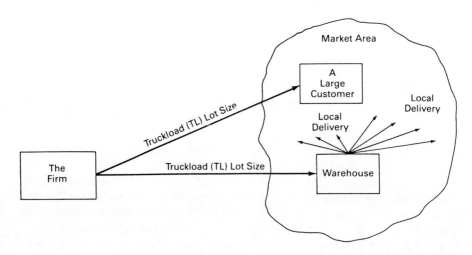

Fig. 9.3 Flexibility in Product Distribution

flexibility of this operation permits least-total-cost transportation while improving customer service.

Postponement

One characteristic of logistics is the movement and storage of items in anticipation of future demand. Material is transported to the firm and stored in anticipation of a need during production. The finished good is moved to the warehouse and stored in anticipation of future sales. Inventory is maintained in anticipation of future need for repair or in response to anticipated consumer demand. These actions, and the concurrent commitment of resources that is incurred, are a necessary part of the logistic system. While necessary, these actions also entail a degree of risk, as anticipated events frequently do not occur as projected.

A method of reducing this risk is to delay or postpone either (1) the final steps in the manufacturing process leading to the product or (2) the logistic activities associated with the product. There are two types of postponement—form and temporal—that reduce or totally eliminate this risk by the early attainment of customer commitments (the product is not completed or the logistic activity is delayed until after the customer makes a commitment to purchase). *Form postponement* is best illustrated by the retail paint store. Paints are available in literally hundreds of colors that vary in popularity with the consumer. Cans of paint are relatively bulky, and an inventory offering a fair degree of availability would require an immense storage capacity. The retail paint store avoids the problem, and enjoys a major reduction in inventory expense, by storing only one color of paint—white. Small tubes of a few colors are then mixed in various proportions and added to white paint to meet consumer demand for each and every color that is identified as available. This postponement, which is a delay in the final manufacturing step, permits a relatively small inventory to meet a customized and wide-ranging consumer demand. *Temporal postponement* is a reduction in transportation and storage requirements as opposed to the delay in manufacturing. This type of postponement involves storing a large assortment of products in a few large warehouses at central locations. Shipment beyond these points is delayed until the customer has made a commitment to purchase. The catalog order store is an excellent example of temporal postponement.

Consolidated Shipments

Another method of reducing logistic cost is by using *consolidated shipments* to reduce transportation charges. The combination of many small shipments into one large shipment may lead to a significantly lower rate per hundredweight. This can be achieved through market area grouping (consolidating small shipments to a specific market area) or through schedule distribution (using a predetermined shipping schedule to build up larger shipments). The potential for consolidation of shipments, while it varies with the firm, the product, and the market, is a valuable addition to the ILS manager's package of logistic resources.

Reverse Logistics

Reverse logistics refers to the backward movement of items through the logistic channel (from the consumer to the manufacturer). This was essentially initiated by the U.S. government by the mandated recall of automobiles thought to have serious safety defects. This specialized form of logistics is becoming more common because of an increased emphasis of quality standards, environmental concerns leading to increased usage of returnable beverage containers, and defined responsibilities for hazardous products. The concept of implied warranties has also impacted the magnitude of this reverse logistical flow. *Implied warranties* say that a product is warranted to perform as intended (the toaster will toast), regardless of any expressed claims that attempt to limit responsibility. As may be imagined, reverse logistics is an expense to the firm and one that cannot be economically justified by potential increases in revenue. The justification of reverse logistics lies in service to society.

SUMMARY

Logistics is a support function, and the product of logistics is performance. Performance within the logistic element is a function of availability, capability, and quality. Availability is a function of safety stock levels and is measured by the probability that an item will be available when requested. Capability refers to the speed and consistency of logistics performance cycles, whereas quality is a function of the number of incorrect, missing, or damaged items. Each of these measures must be taken over time to preclude erroneous conclusions regarding the level of logistics performance.

Logistics performance is based on and implemented for the customer. The end user or final customer is the consumer who purchases the product. This customer is the ultimate determinant of the need for logistics. Logistics, however, exists to provide a service, and the customers of logistics include all who have need or and request the service. This definition of the customer incorporates the consumer, individuals and functions within the firm, and individuals and functions operating within the environment.

Improvements in service to the customer may be implemented through changes in the structure of the logistic system. Variations in safety stock levels decrease stock-outs, thereby increasing inventory availability. The performance cycle may also be changed by either increasing the number of logistic facilities or by selecting alternate transport modes. Changes in the structure of the logistic system are expensive and their economic justification depends on an increase in sales as a result of improvements in customer service. The anticipated increase in business is a projection of what *may* occur. Thus there is a degree of risk present in all decisions to change the level of customer service. Management must properly evaluate that risk and make the final decision.

Operating policies in the firm offer alternatives to changing customer service levels through changes in the structure of the logistic system. These approaches to lowering costs include the addition of flexibility to the firm, form and temporal postponement, and consolidated shipping operations.

Reverse logistics, or a flow through the logistic channel from the consumer back to the manufacturer, is a cost to the firm that is justified as a necessary service to society.

QUESTIONS FOR REVIEW

1. Why is logistics described as a service function?
2. Describe the concept of availability.
3. What is used to measure the logistics performance cycle?
4. Describe two methods that may be used to vary the logistics performance cycle.
5. Define the customer.
6. How may the firm add flexibility to its operations?
7. What is the difference between form postponement and temporal postponement?
8. Describe the process of determining if increased levels of customer service are economically justified.
9. What are consolidated shipments, and how can they benefit the firm?
10. What is reverse logistics?

10

PERSONNEL AND TRAINING

Training differs from education in that it is task-oriented and specific, whereas education is concept-oriented and more general in nature. The objective of training is to learn the "how to" of an activity specifically dedicated to a defined purpose. This objective is not intended to educate the trainee in the sense that the university system provides an education. It is true, however, that a limited amount of education is certainly a byproduct of training. The differences between the two are not clear-cut and distinct; however, both involve the transfer of knowledge and skills from the trainer (or professor) to the trainee (or student). Training places the greater emphasis upon the acquisition of skills, whereas education emphasizes the acquisition of knowledge. The methodologies used in the process are, however, very similar, and the "good" trainer can function with equal facility in either environment.

Training as an Element of Logistics

Training, as an element of logistics, provides a service to the firm in the same manner as any of the other logistic elements. It may be viewed as a system having an input, a transformation process, and an output just as the firm and the functions within the firm are viewed as a system. The training system is illustrated in Figure 10.1. Incoming personnel who lack the knowledge and skills necessary to accomplish some activity are inputs to the training system. The transformation process is

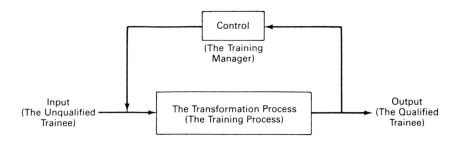

Fig. 10.1. Training as a System

the transfer of skills and knowledge that takes place between the trainer and the trainee, and the output is personnel possessing newly acquired skills and knowledge.

Training may also be considered within the context of the logistics performance cycle. This concept is illustrated in Figure 10.2. The *training performance cycle* is characterized by a high degree of variation. A knowledgeable staff of instructors familiar with the subject matter and course content may lead to a performance cycle duration that is approximately equal to the training course length. On the other hand, a new training program or a less experienced instructor may require an extended period of time for course development and instructor preparation. This may lead to a performance cycle duration that is ten to twenty times the duration of the actual presentation. Training course lengths that vary from a few hours to several weeks also contribute to the variability of the training performance cycle.

The Logistics Instructor

The logistics instructor is the trainer who provides a training service to the human element of the integrated logistic support arena. The instructor responds to a service request by developing and presenting a program of instruction in a logistics-oriented training course. This process is no different from that of a professor developing a course in, for example, behavioral science and management in response to a request from the university. There is a difference, however, in that many courses presented by the logistics instructor have not been presented before and may not be presented again (by the logistics practitioner).

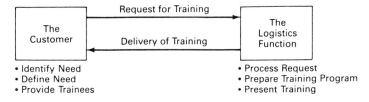

Fig. 10.2. The Training Performance Cycle

The logistics instructor represents the incorporation of a wide range of skills and knowledge into a single dynamic resource. Products of the firm typically address a multiplicity of disciplines, and the instructor must be proficient in all of them. The home video recorder, for example, is a combination of electronic and mechanical technologies. Military systems represent an even greater diversity because a single system may include elements of electrical, electronic, digital, optical, and mechanical technologies.

The typical firm does not have the resources needed to maintain an instructor staff in each of these disciplines. As a result, the logistics instructor must "wear several hats" and possess the ability to function with equal facility in many areas. This problem is simplified somewhat in that the instructor has access to a wide range of engineering talents within the firm.

Logistics Instruction

Instruction provided as a service of the logistic function of the firm is not directed toward performance within the logistic element. The transportation manager who requires a knowledge of linear programming for the evaluation of alternative routings in product distribution must acquire this knowledge in the academic environment. Logistics instruction and the logistics instructor represent an activity in the logistics arena. It does not provide training (or education) in the practice of logistics.

The instruction provided by the logistics function is directed toward the personnel who operate and maintain the products of production. It should be recognized, however, that this may also include personnel within the firm as would be the case when the products of production of one firm represent the equipment or production to another firm.

Advances in technology and increasing product complexity have resulted in an increased awareness of the importance of training. The ever-increasing sophistication of consumer products such as the personal computer requires that even casual users receive a few hours of instruction. The repair of these items require much longer periods of training if the product is to be adequately supported following transfer of ownership.

The problem is even more acute in the firms that produce products for the U.S. government. Large and extremely complex systems produced for the military are at the forefront of technology. These are new systems requiring extensive operation and maintenance training if they are to successfully perform intended missions.

The Logistics Training Course

Logistics training courses may be a few hours, a few weeks, or a few months. They cover an incredible diversity of subjects, limited only by the creativity inspired by technology within the firm. Regardless of training course lengths or the diversity of course content, the final product is derived through the application of methodologies common to all techniques used in the transfer of knowledge among individuals.

Logistics training courses exist for the purpose of developing a specific skill that the trainee requires in meeting defined objectives. Skill development implies a degree of proficiency in the performance of an activity. Proficiency in such manipulative skills cannot be gained in the typical classroom setting; therefore, logistics training is characterized by a high degree of trainee involvement. The logistics course addresses this by dedicating a minimum of 50 percent of total course time to trainee interaction with the product or equipment being presented.

Logistics training courses are not, however, limited to a combination of classroom presentations and practical exercises involving trainee and equipment interaction. To the contrary, the classroom portion of the training may be supplemented or replaced by a variety of instructional methodologies such as self-study, self-paced instruction, video taped training programs, computer-aided instruction, or other methods that have been developed to support the learning process. Trainee participation, however, must involve the student and the equipment (or a specifically designed training aid that accurately simulates the equipment) in realistic scenarios.

Prerequisites to Training

Logistics training begins with the request for service that describes, in general terms, the type and nature of the training needed. Processing this request entails an evaluation of the potential student population and of resident capabilities to determine if the logistics training function possesses the resources necessary to comply with the request. If the necessary instructional resources are not available, the logistics training group must (1) borrow the resources from other functions within the firm, (2) obtain the resources from the environment of the firm (through new hires or a subcontract with another firm possessing the needed talent, or (3) turn down the request for service because of a lack of capability.

The last choice is by far the least desirable approach, since the purpose of logistics is to provide a service. Additionally, the customer who is refused service will seek it elsewhere, thereby lessening the value of logistics to the firm. The first and second choices lead to accomplishment of the task.

Types of Training

Logistics training courses are loosely grouped under the heading of operation or maintenance training. This is an overly simplistic description as, for example, there is considerable difference between training to operate a video recorder with a camera mounted in a single unit and a complex system developed for the U.S. Navy.

Maintenance training is also highly variable, since it may address any one of the three maintenance levels (organizational, intermediate, or depot). The course content and the length of training would significantly differ for each of the three maintenance levels.

Student Population

Training, if it is to attain a common level of achievement, must be presented to a body of students having a degree of commonality in prior skills, knowledge, and education. A diversity of backgrounds virtually excludes the presentation of a successful training program. Therefore, the logistics instructor must develop an understanding of the potential student population prior to any attempt to develop the planned course of instruction. The type of training requested provides an initial indication of potential student capabilities. For example, the retail merchant may request training for the sales staff to permit knowledgeable product discussions with potential customers. This training would emphasize the operational capabilities of the product, thereby permitting demonstrations of the product's value to the consumer. The customer service representative, on the other hand, may require training in product maintenance and repair.

The nature of the product indicates the type of student that may be expected. It is a relatively safe assumption that the customer who requests maintenance training for a computer will provide students who possess a knowledge of digital circuits and computer technology. Training on systems produced for the military services also indicate probable student capabilities. The instructor can determine with very little effort the specialty assigned to the system. The military technical training received by the potential student can then be reviewed. The job specialty and the requisite training leading to that specialty provide all information needed to determine probable capabilities of the incoming student.

A Training Philosophy

Logistics training provides the student with an identified degree of proficiency in the accomplishment of task-oriented objectives. The critical nature of this training is evidenced by reliability studies that show that as much as 60 percent of product or system failures can be attributed to human errors of either commission or omission. Training, in attempting to improve on this deplorable statistic, must address the functions that are the province of the human element. In general, these functions consist of the following:

A. Receiving information from the system and providing information to the system. These tasks involve data recognition and response to stimuli by the operator or maintainer.
B. Making decisions through information processing in order to predict, extrapolate, transfer functions, and estimate conditions.
C. Performing some action involving motor and anthropometric capabilities.

Product and system design has a significant impact on the training process through the allocation of functions between the hardware and software element and the human element. Ideally, this allocation should combine to provide the most reliable and efficient human-machine interface. Unfortunately, human engineering and product maintainability are not always considered to the extent necessary to achieve this goal. The resulting less-than-optimal configuration increases the cost of the human element and increases the difficulty of training.

Training Objective

The objective of logistics training is, quite simply, to provide the student with new capabilities. The critical first step, as previously discussed, is to evaluate pre-existing student capabilities. The second step is to evaluate the product or system being presented to determine the capabilities the student must acquire to meet training program objectives, that is, the capabilities the students must have upon completion of the training. The difference between the capabilities that exist prior to training and the desired capabilities upon completion of training define what must take place during the transformation process of the training system. It should be apparent that the absence of a difference would negate the requirement for training.

Steps one and two, therefore, serve to identify what is required by the training course. Each task or activity that has been defined as being necessary and that is not included in the current capabilities of the student represents a training requirement. The training requirements include only those activities needed by the product or system in question. Additional material such as "nice-to-know" information and detailed theory are specifically excluded from this training. Since optional material is excluded, the student is required to pass each segment of the training program. The course only includes material that is required to operate or maintain a product or system. If the student can fail one segment of instruction and still meet the training objectives, the failed segment is not needed in order to fulfill job performance requirements. If it is not needed, the segment should be omitted.

Training objectives include the total course objective and objectives for each instructional segment or lesson included in the course. Course objectives are typically more general in nature, since they specify the overall purpose of the training program. Course objectives are also referred to as *terminal objectives* because they describe the capabilities the students must have at the completion of training.

Course objectives are supplemented by objectives that describe the purpose or intent of each lesson. Lesson objectives define the capabilities acquired by the student as a result of each instructional segment. The aggregate of all lesson objectives must equate to the course objective; thus they are also called *enabling objectives*.

Writing the Objective

Training objectives, both terminal and enabling, express the skill or knowledge the student is expected to acquire in the training. Moreover, the objective addresses skills and knowledge that can be quantified and objectively measured. However, a simple statement such as "at the conclusion of this training the student must be able to operate the Brand X video recorder" is not a good objective. The good objective addresses (1) the activity that is to be performed, (2) the conditions under which the activity is to be performed, and (3) the standards or criteria for measurement of the activity. An example of this three-part objective is as follows: (A) At the conclusion of this training the student must be able to successfully operate the Brand X video recorder, (B) given an operating Brand X video recorder and television monitor, a source of power, interconnecting cables, required tools, and the operating instructions provided with the recorder. (C) Successful accomplishment of this objective shall be determined by, (1) the student connecting the video recorder to the television monitor, (2) recording a short segment from each of five different television channels as selected by the instructor, and (3) playback of each recorded channel. All activities are to be completed in no more than twenty minutes.

This example, while admittedly elaborate for the purpose of illustration, demonstrates the three elements of a good objective. It defines a specific action that is expected to occur as a result of training; it accurately defines the conditions for accomplishment of the action; it establishes objective standards for measuring successful accomplishment.

A common failure in the preparation of objectives is using statements such as "To explain the operation of the abc system." This, even if it is expressed as a three-part objective, is an activity of the instructor. Objectives are directed toward the student and what he or she is expected to gain from the training.

Training Requirements Analysis

The analysis of training requirements is a function of the training course objective. Training in product or system operation depends upon planned utilization and operating scenarios, whereas maintenance training relies heavily upon reliability, availability, and maintainability (RAM) analysis. This is not, however, a clear-cut distinction because a maintenance capability also requires a knowledge of operating techniques and procedures.

RAM analysis, with the assistance of other disciplines such as system and test engineering, provides the information necessary to determine

- A. The parts that are likely to fail
- B. The requirements for tools test and support equipment
- C. The maintenance tasks and the estimated times required for those tasks
- D. The features, techniques, and procedures of testing

These data are used by the training staff to evaluate the operation and maintenance tasks and divide them into related task groupings and skill-level categories. The resulting matrix identifies the tasks that must be performed by operation and maintenance personnel and the requisite skill levels that are necessary for successful accomplishment of product or system responsibilities. These are the capabilities the student must possess upon completion of training. The projected capabilities of the student candidate are now compared with the identified job requirements. The difference between the two levels of capability define what must be provided through training. This concept is illustrated in Figure 10.3.

The identification of job tasks and the associated performance methodology is one of the most difficult challenges facing the training staff. The best source of information concerning job performance requirements is, of course, the experienced individual who is performing in that capacity. The job incumbent may, in fact, sometimes be the only source of information for determining course requirements. Several methods involving various survey techniques and procedures may also be available for gathering this information. These methods include the following:

A. *The occupational survey.* Occupational surveys obtain data about job duties and the tasks that are actually performed from a large number of job incumbents. These data are then tabulated by computer in the form of an occupational survey report that reveals all tasks performed in each job specialty.

B. *The questionnaire.* Questionnaires represent a method of obtaining job performance information through mail surveys. Job incumbents are asked to provide certain background information and to describe job duties and responsibilities. The descriptions include all tasks that are performed and any tools or equipment that may be required. Job incumbents should prepare the questionnaire independently, although the completed data should be reviewed by a supervisor to add comprehensiveness and validity to the information being provided. The chief advantage of this method is economy. Major disadvantages include the time required to obtain information and the typically low response to mail surveys.

C. *The checklist.* Checklists differ from questionnaires in that checklists include listings that are believed to describe the job task. The incumbent is asked to indicate, by means of a checkmark, the duties and tasks that are actually performed. The result is a job inventory for that position. The primary advantage of this method is the relative ease with which it can be administered to large groups. It suffers, however, from the difficulty of constructing checklists that clearly communicate the exact nature of job duties.

D. *The individual interview.* This method relies upon the selection of representative workers for in-depth interviews concerning duties and tasks that are performed. Interviews are conducted individually, usually at a location removed from the work site, with the information recorded on standard forms. The combined results of all interviews are consolidated in the job inventory.

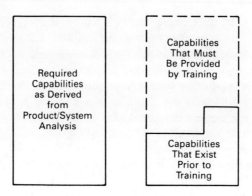

Fig. 10.3. Identification of Training Needs

This method typically results in the highest quality data of all the other methods. It is, however, time-consuming and relatively expensive.

E. *Observation.* This method is similar to the interview except that it must take place at the work site. The incumbent is questioned about the work being accomplished as duties and tasks are performed. Observation yields information that is more complete and accurate than that obtained through other methods. The disadvantages are its slowness and the necessary interference with work routines.

Job-task analysis is difficult, time-consuming, and expensive. It must be performed, however, as training is effective only to the extent that it provides personnel qualified for the job. This necessitates an inventory of the duties and tasks that are to be performed by the human element. These tasks then become the basis for training. Thus training requirements, as identified by the difference between existing capabilities and desired capabilities, establishes course content. The next step is to derive an estimate of the optimum course length.

Course Length Determinants

The length of the training course is a function of course content and the tasks the student must successfully accomplish in bridging the gap between actual capability and desired capability. Tasks and the time required for their completion provide a means of approximating the optimum length of a course.

Determining task lengths or the time interval needed for task performance are accomplished through (1) estimates based upon the times required for similar tasks, (2) discussions with individuals engaged in similar activities, and (3) consultations with maintenance (maintainability) engineering. Individual time increments for each task are then totaled to determine the total task performance time that is required. Task times for the tasks that are already a part of the prospective student's current capabilities are now subtracted from this total time. Current capabilities are excluded from further analysis since there is no need to provide training in these areas. The tasks that remain are those that must be acquired.

Individual task times represent the time the skilled worker needs to perform the task. Students who are developing a performance capability require more time to perform the same task. Aggregate task times should therefore be multiplied by a factor of six to accommodate the learning environment. This now becomes the allocated time interval for hands-on training wherein the student learns through interaction with the product or system. Hands-on training, however, represents only about one-half of a performance-oriented training program. The other half is time for instructor presentation, demonstrations, and related learning activities. Hands-on time is therefore multiplied by two to arrive at the total course length.

Figure 10.4 illustrates the process used to determine course length. The example course of Figure 10.4 has a total duration of seventy-two hours or nine days assuming 8 hours of instruction per day. A more typical example would limit the instructional day to 6 hours resulting in a total course length of 12 days. This length approximates the ideal for the performance-oriented training program depicted in the example. Course lengths are, however, subject to variations because of budget constraints, the availability of student personnel, access to equipment, and schedule restrictions. The actual course length may vary considerably from this model. The real value of this model is to identify what *should* be and is a reference point for needed changes and perturbations.

Time Phasing

Successful implementation of the training course demands proper time phasing with respect to other elements of the program. Training requirements analysis, for example, cannot be accomplished until maintenance engineering has identified the tasks and the times required for task performance.

A primary impediment to scheduling of training courses is the availability of an operating product or system. A lack of equipment precludes hands-on training, thereby severely limiting the value of a performance-oriented training program.

Task	Time in Minutes
1	18
2	24
3	17
4	31
5	33
6	14
7	65
8	47
9	28
10	76

Σ All tasks × Learning factor = Hands-on time
6 hours × 6 = 36 Hours

Hands-on time × 2 = Total course
36 × 2 = 72 hours or 9 days' training

Σ All tasks = 353 minutes (approximately 6 hours)

Fig. 10.4. Course Length Determination

Other factors that must be considered in establishing a schedule for training include the availability of tools, test equipment, and technical publications relating to operation and maintenance of the product.

Preparing the Training Program

Training requirements analysis defines the specific skills that must be acquired and, through extrapolation, probable course lengths. The next step is to develop a training program that will enable the student to effectively and efficiently attain the skills and knowledge necessary for the performance of required activities. This development begins with a thorough understanding of the subject matter and material that should be included in the training course.

Scope of Training

The purpose of logistics training is to meet customer requirements. In defining those requirements, an analysis of the object of training (the product, piece of equipment, or system) is combined with an assumed student population to determine probable course content and course length. This description of the overall course is then followed by specific definitions leading to the design and development of a viable training program. As an example, assume that a firm has developed a large data base management system (DBMS). Sales have resulted in customer requests for the firm to provide DBMS user training for customer personnel who will use the product.

To understand the planned DBMS user training, one must understand the purpose that the planned training is intended to serve and the type of characteristics the prospective student should have. For instance, should it be expected that the student candidate possess (or gain through training) a knowledge of the various technical aspects of the product he or she is to use? Should the prospective student become intimately familiar with the detailed operation and functioning of the product as it is being used in fulfilling its designed purpose? The answer to these questions is an emphatic No! Logistics training is specifically intended to present only the information necessary to perform an activity. Material not needed in job performance should not be presented.

A user, within the context of this example, is an individual charged with the utilization of a device, product, or system in meeting a specific objective. Consider the driver of an automobile. The driver (user) utilizes the automobile (system) to attain the objective of movement between successive points. In using the automobile, the user does not need to understand the inner workings of the engine, drive train, or other mechanisms that permit the system to function. In a similar manner, the DBMS user does not need to understand the inner workings of the DBMS. It is simply a tool that provides the resources needed in meeting an objective.

The objective of using the DBMS, again within the context of this example, will differ among various users. An example of the diverse objectives among users may be explored by considering a different form of data base—the repository of knowledge known as an *encyclopedia*. A user possessing a background in integrated logistics support management may use this vast resource in obtaining information upon which to base a decision. The information thus sought would, however, be very different from that desired by the engineering student. The important point is that a common resource (the encyclopedia) is used by various individuals to meet different needs (objectives). It should be recognized, however, that while the information solicited from the resource varies, the process or methodology for extracting the information is virtually identical.

Users of the hypothetical DBMS also have multiple requirements. The process of accessing the DBMS to meet those requirements is, however, the same for each requirement. Thus the assumed training, in its simplest form, should provide instruction in (1) what is available from the DBMS and (2) how to access that resource to meet the specific objectives of the individual user. This definition limits the scope of the planned training course to what is needed in meeting a defined objective of how to use the DBMS. The instructor staff can now proceed with course development.

Course Prerequisites

Course development is designed to impart a specific range of skills and knowledge. This development assumes some minimum entry level for students attending the training. The entry level defines prerequisite skills and knowledge, thereby permitting each student to begin training at approximately the same knowledge base.

Entrance criteria is only one factor in the development of a successful training program. Equally important factors include the selection of a knowledgeable and capable instructor staff and the instructor-student ratio. Performance-oriented training involves a high degree of student-equipment interaction in actual or simulated exercises. This type of instruction dictates a relatively low instructor-student ratio to permit each student sufficient time to practice the required activities. The ideal is, of course, one to one. This, however, is not cost effective and is rarely seen. A more realistic figure is four to five students per instructor. This again may vary significantly, primarily due to budgetary constraints, although the schedule for training and the number of students to be trained may also have an impact.

Preparing the Instructor

Instructor preparation involves the activities associated with learning the product or system to the degree necessary for presentation of an acceptable training course. The primary method of accomplishment is through study and access to resources within the firm. Resources within the firm include the engineering staff responsible for product development and others who possess an in-depth knowledge of the item.

The product of the firm frequently incorporates assemblies or equipment that are products of other firms. The instructor may be required to attend a training course offered by another firm. This may, at first glance, appear expensive. However, it could be very cost effective in that learning through study of technical publications may require a much longer period of time.

The time required for instructor preparation is highly variable in that it is a function of the experience and knowledge of the selected instructor. The instructor possessing a thorough knowledge of the product may require little or no advance preparation. The less knowledgeable instructor will, of course, require a significantly longer preparation interval. The average preparation time for a combined classroom and on-hands training course is five to six hours for each hour of instruction. This does not include the time required to develop supporting course materials.

Supporting Course Materials

Logistics training courses are oriented toward products of the firm and are provided at the request of customers. These training courses typically include students who are qualified instructors from the customers' internal resources. Their purpose in attending is to develop an ability to present subsequent training courses through the use of internal resources. As the training course is destined to be included within the customer's inventory, the request for training usually includes a stipulation that supporting course materials be prepared in a format preferred by the customer. There are numerous exceptions to this, however, as it is usually less expensive if the firm presenting the training uses its internal standards. Variations in the format and style of supporting course materials preclude definitive statements regarding each item. There are, however, many features that are common to these materials regardless of the specified format. The general content and purpose of support materials are addressed in the following paragraphs.

Training Plan

Training plans provide the customer with information regarding the methods the firm intends to use in the training. This document serves a dual purpose: It forces the firm to plan and organize the training program and provides the customer with an opportunity to review the plans of the firm to verify that the program does, in fact, meet the identified need. Training plans are typically arranged in three distinct parts or sections.

The first part of the training plan presents the firm's approach to meeting the need for training and the methodologies to be used in responding to this need. It includes such training-related factors as the course objective, course length, course schedule, student prerequisites, instructional methodologies, support expected from the customer, and a preliminary outline of the course. The purpose of this part of the plan is to assure a mutual understanding of both the need and the method to be used in satisfying that need.

Frequently the request for training received from the customer is not specific or it can be resolved through alternate approaches. Thus the plan should present recommendations of the firm regarding each alternate approach to resolving the expressed need. The customer can then select the preferred method or solution.

The second part of the training plan is devoted to the capabilities of the firm in responding to a request for training. It addresses factors such as the organization of the training function within the firm, qualifications of the resident instructor staff, and the training facilities that are available to support training. The purpose of this section is to provide the customer with a summary review of the firm's credentials.

The third part consists of recommendations by the firm regarding continued support to long-term training needs of the customer. This section represents an attempt by the firm to attain a part of the market by becoming the training resource for that customer.

Course Outline

Course outlines provide a summary description of the training course. It includes an updated version of the training course description included in the first part of the training plan. The main portion of this document consists of a detailed and sequential listing of each major topic that is to be included in the course presentation. Training course outlines define the course in detail and provide a reference for use by the instructor staff during subsequent course development activities. The course outline is also referred to as the *program of instruction*.

Lesson Guides

Lesson guides are detailed documents that lead the instructor through the course. A lesson guide is prepared for each major topic identified in the course outline with subtopics, as necessary, to assure a complete and thorough presentation of needed material. Each lesson guide includes the objective for the lesson, the instructional methodology to be employed, the time allotted for the lesson, and references to source data used in developing the lesson. Lesson guides should provide all the information that is needed for training course development, thereby permitting an experienced instructor to prepare for and present subsequent courses of instruction.

Hands-on Training Guides

The hands-on training guide is a guide for the student to use during hands-on training. This guide restates the lesson objective and provides the student with an identification of any required tools, test equipment, and support materials that are needed in performing the indicated activity. Hands-on training guides emphasize safety techniques and procedures that are applicable to the lesson and include space for student responses to events that occur during the training activity.

Audio-Visual Aids

Audio-visual aids encompass a wide range of media that may be used to enhance the learning process. These items may consist of audio recordings, video recordings, overhead projections, slides, charts, or any other devices that are appropriate to the situation. The important feature of audio-visual aids is they provide equal access to each student in the classroom environment.

Student Text Material

Student text material includes copies of related technical publications or appropriate extracts thereof. For example, a class limited to the operation of a specific product would have no need for maintenance and repair information. The student text would consist of the portion of the publication that relates to operating procedures and descriptions of operating controls and indicators.

Examination

Training involves the transfer of knowledge and the development of skills. The success of training can only be verified through an evaluation of student capabilities resulting from this training. Examinations are the vehicle for this evaluation.

Logistics instruction is directed toward the attainment of a performance capability, and the examination should pursue the same goal. Although the written examination is certainly not excluded, the emphasis is on student performance of activities learned throughout the course.

SUMMARY

Training provides a service to the firm in the same manner as any of the other logistic elements. It is a system having an input (the untrained student), a transformation process (the learning situation), and an output (the trained student). Logistics training is performance-oriented and specific as opposed to education, which is concept-oriented and general.

The term *logistics training* does not refer to training the individual in the logistics arena. Instead, logistics training is provided by the logistics function of the firm to customers of the firm and is normally related to products produced by the firm.

A logistics training course is initiated by a customer request for service. A necessary first step in responding to a request is to determine the type of course required by analyzing the tasks and activities necessary to use the object of instruction. Task analysis is accomplished in a variety of ways, including (1) the occupational survey, (2) the questionnaire, (3) checklists, (4) individual interviews, and

(5) observation. The resulting inventory of job tasks defines the activities that occur in operational or maintenance support of the item under consideration. This defines the capabilities that a user must have at the completion of training.

This information is then compared with the presumed capabilities of the student population prior to training. The difference between existing capability and required capability is the instruction to be provided through training.

The performance-oriented aspect of logistics training emphasizes training in only what is necessary for task performance. "Nice to know" information and related data are specifically excluded as not necessary and therefore not needed.

Logistics training is normally offered to a customer just once. The customer then trains its own personnel through its own internal resources by an instructor who attended the logistics training course. Training support materials are also made available to the customer following course presentation. Training support materials include (1) the training plan, (2) the course outline of instruction, (3) lesson guides, (4) hands-on training guides, (5) audio-visual aids, (6) student text material, and (7) examinations.

QUESTIONS FOR REVIEW

1. Describe the training system.
2. What is the training performance cycle?
3. What is the difference between training and education? What are the similarities?
4. Describe a method of determining probable capabilities of the prospective student.
5. What is the primary factor leading to a determination of what should be included in the course of instruction?
6. Why can it be said that the student must pass each element of the logistics training course?
7. Describe a well-written training course objective.
8. What is the purpose of the training plan?
9. Describe the preferred method of determining job and task performance information.
10. What is a performance-oriented examination? Why is this type of examination preferred for logistics training courses?

11

TECHNICAL PUBLICATIONS

Technical publications incorporate a vast array of material designed for the expressed purpose of providing information about a product. The purpose of technical publications is to inform. The term *technical publication* is not limited to a book or manual comprised of the printed page. Rather, it refers to all media (microfilm, video tape, computer displays, and so forth) that serve to inform. The well-prepared technical publication is tailored to the needs of a defined audience: It provides that audience with assistance in the performance of specific tasks or in meeting assigned objectives and permits the inexperienced individual to perform on a par with the skilled practitioner.

A lesser quality publication, on the other hand, has the opposite effect: It is avoided by the user because it is confusing, lacks needed detail, is difficult to understand, is poorly organized, or is deficient in other ways. This leads to a reliance on extrapolation from prior experience, information gained from others, or an expensive and time-consuming trial-and-error approach. These methods all too frequently lead to unnecessary damage or destruction of the product, item, or system purportedly being supported by the publication.

Technical publications range from pamphlets that describe the use of a consumer product such as an electric clothes drier to a multivolume library. An example of the latter is the technical publication that describes all aspects of the operation and maintenance procedures necessary for the successful operation of complex systems procured by the U.S. government. Regardless of the level or magnitude of content, the well-prepared and user-oriented publication should enable the user to attain a certain capability or improve an area of performance, resulting in optimum use of the system to accomplish its objective, whether the objective is to dry clothes or perform a complex mission for the Department of Defense.

The Technical Writer

The author of technical publications is the technical writer. Technical writing is a vastly different profession from that of other writers in that the technical writer is denied freedom of expression. The profession is dedicated to detailed descriptions of products and their characteristics. This leaves little room for the exercise of creativity. The technical writer can only respond to the product and what it does. Writers for the popular market, on the other hand, depend upon the reader for product demand. The writer must respond to the desires of the marketplace or select an alternate line of business. The technical writer is not faced with this constraint. The audience for this writing also has no alternative choice and thus cannot seek another outlet for the information. Recognizing this, there is a tendency for some technical writers to write for themselves, thereby ignoring the needs of the reader. Denied the creative outlet, they seek solace in elaborate technical verbosity and obtuse terminology. This unfortunate attitude is changing, however, in line with the market-oriented philosophy of business. Technical writers are realizing that they have a responsibility toward their audience.

The term *technical writer* is, in many ways, a misnomer. Virtually everyone in business or industry writes, and the majority of this writing is technically oriented. But are they all technical writers? Most assuredly not, as their writing, with few exceptions, does not involve the transfer of knowledge (communication). A more appropriate term for the technical writer would be *technical communicator,* since their objective is to communicate. Technical information possessed by the writer must be transferred to the reader, thereby becoming the common property of both. This transformation process involving the communication of knowledge and understanding between writer and reader is the sole objective of technical writing.

The Technical Publication

Writing, whether technical or nontechnical, is judged or evaluated on two levels: (1) the level of literacy and (2) the level of competence. *Literacy* incorporates the rules of grammar, language usage, and punctuation. The writer must possess an extensive knowledge of these factors to permit understanding by the reader and to reduce or eliminate distorted communications between writer and reader. Literacy of the presentation must, however, take second place to the writer's level of competence or ability to transfer knowledge and understanding. *Level of competence* is related to factors such as the clarity of presentation, organization, accuracy, adequacy, relevance, and effectiveness.

Clarity of Presentation

Clarity of presentation is related to the writer's style and skill as an author. Technical publications support products of the firm and are prepared for a diverse audience. The writer must recognize that this audience varies widely in educational background, knowledge, and skill level. The publication must be prepared to a level of understanding that corresponds with the intended audience.

Clarity of presentation is similar to the concept of readability and represents one of the most difficult aspects of technical writing. Readability in writing is a function of sentence length, the number of multisyllable words, and other factors. It results from both educational achievements and career specialties. For example, the physician who is writing for other physicians would (or should) write in a different manner than the physician writing for the general public. Looking at it from a slightly different perspective, the physician possesses a certain reading level in the field of medicine. This level may be quite different for the physician reading in an unrelated field such as astronomy.

The difficulty in writing for clarity (readability) lies in the relationship between the writer and his or her subject. The writer must possess a great depth of knowledge regarding the object of the technical publication. This knowledge may, however, need to be expressed in prose to a reader with a lesser degree of knowledge or a lower reading level. It is very hard to "write down," because the natural tendency of the author is to consider such writing as an unfavorable reflection on his or her educational achievements. In fact, a great many technical writers tend to "write up" for this very same reason.

Organization

Organization refers to the structure and layout of the technical publication. Material in the publication should be arranged in an orderly and logical manner to assist the reader in (1) the development of new skills and knowledge or (2) the enhancement of existing skills and knowledge. User-oriented documents such as the technical publication should be structured to facilitate occasional use as reference or source material. This permits efficient use of the publication as the skills and knowledge of the reader increase.

Accuracy

Accuracy defines the element of realism that exists within the publication. Publication content must be thoroughly checked and verified against the subject of the publication. Even relatively insignificant errors or omissions severely impact the credibility of the publication, thereby limiting its usefulness and ability to meet intended objectives.

Publication content may be verified by technical review cycles, wherein the material is subjected to reviews by other functions within the firm where the necessary knowledge of the product may exist. This review must be performed by an individual or individuals other than the author. Additionally, procedural data should be verified through actual performance. An excellent means of verifying that the publication meets intended objectives is to have it reviewed by individuals possessing skills similar to the intended audience. This type of review includes both reading for understanding and the performance of procedural activities.

Adequacy, Relevance, and Effectiveness

Adequacy, relevance, and effectiveness are all characteristics of the publication that enables the reader to attain specific and implied skills and objectives. The publication is developed and prepared to fulfill a specific objective or purpose, and the contents must include information that serves that purpose while excluding extraneous and gratuitous material. For example, the publication received with a new automobile is designed to acquaint the new owner with the characteristics of vehicle operation. This purpose requires information on operating controls and indicators, starting the vehicle, driving characteristics, and similar data. A presentation on the theory of internal combustion engines is neither needed nor necessary in meeting the objective of the publication.

The Intended Audience

Who are the prospective readers of the publication? This important question must be answered before beginning the writing task. The question becomes even more critical when the subject matter is complex and difficult to understand, which is the case with the majority of technical writing efforts. One way to define and characterize the intended audience is to evaluate the readers' probable educational level. This information indicates the writing style and type of vocabulary to use in preparing the publication. This analysis, although seemingly difficult, is relatively simple. In many cases the product itself provides a realistic indication of the user's probable educational background.

For example, consider three unrelated products: the electric toaster, the personal computer, and a complex electronic system developed for the U.S. government. An electric toaster is found in virtually all U.S. households. Product operation is relatively unsophisticated, and users of this appliance span the spectrum of society. Individual educational backgrounds in this large population can range from the high school dropout to the advanced postgraduate. While the highly educated consumer is capable of comprehending a lower level writing style, the converse is not true. Therefore, the publication describing the operation of the electric toaster should be written so that it is easily understood by individuals possessing a seventh- or eighth-grade reading level. Personal computers, on the other hand, are generally the exclusive province of the more affluent and better educated. Operating instructions for this product may safely incorporate a writing style that assumes some amount of education beyond high school. Complex systems developed for the U.S. government represent a special challenge to the technical writer. Yet, they also permit a realistic determination of probable educational achievement. Assume that a system is being developed for a branch of the armed forces. The writer can, with relatively little research, determine the specific job specialties of potential users of the system. The job specialties, in turn, indicate the type of military training courses the prospective users may have completed. A brief review of these training courses and the academic prerequisites established by the branch of service (customer) provides enough information concerning the potential users' expected educational backgrounds.

The Scope of the Publication

The type and scope of information to include in the publication must next be determined. Since it is not possible for a publication to include all knowledge on a subject, it is necessary to select a portion of the subject matter for inclusion in the document. The selected content is, in general, determined by the intended purpose of the publication. For example, an operating guide includes information on operating procedures and the associated controls and indicators. Prose descriptions are supported by clear illustrations of each applicable control or indicator. This type of publication excludes detailed maintenance and repair procedures. The tutorial publication, on the other hand, presents the material in an organized and logical sequence, which supports the learning process. It incorporates numerous examples and reader-oriented exercises designed to assist the user to attain new knowledge or skills.

There are, of course, numerous other types of publications, such as reference manuals, parts catalogs, the illustrated-parts breakdown, computer software, and user manuals. Each type of publication requires a slightly different writing style and variation of content. Regardless of the type of publication, the scope of the contents must be determined *prior to writing*.

A Necessary Prerequisite

Understanding the intended audience and determining publication scope are followed by research to gather information about the subject of the document. This is generally the most time-consuming portion of the task. There is a natural tendency to rush through this activity and a strong temptation to begin putting words on paper as soon as a basic knowledge of the subject is attained. This is a serious mistake. Writing from an inadequate base of knowledge can only lead to an inadequate final product. The publication is inferior, and the writer's obligation to the reader has been compromised.

Publication content is determined by the scope of the document. The scope also determines the subset of the complete range of product support activities that should be included in the publication. The full range of product support includes all activities necessary for the establishment of an operation and maintenance capability. The operation and maintenance capability may, however, be subdivided into a variety of categories, such as

A. Unpacking and installation
B. Turn-on and operational checkout
C. Alignments and adjustments
D. Preventive and corrective maintenance
E. Three maintenance levels
F. Troubleshooting techniques and procedures
G. Packing for shipment

Publication content may include any combination of these and other factors as necessary in meeting defined objectives. Selection of the appropriate subset results in a discrete set of job activities that must be included in the publication. The next step is to analyze the jobs involved in those activities.

Job Analysis

Job analysis addresses the needs of technical writers, who require "what," "why," and "how" information to develop and prepare the technical publication. It also addresses the needs of the user who must read the publication in order to operate or maintain the product.

Technical writers sometimes regard job analysis as simply another demanding job that adds work without providing benefits or usable results. This, however, is not the case. Job analysis, when properly carried out, assists the writer in planning the publication and helps develop the optimum structure of the completed publication. The writer possessing a knowledge of requisite job tasks can easily determine the level of detail necessary for users of the product. Job analysis benefits the technical writer as follows:

A. It provides the writer with all necessary information.
B. It leads to the development of superior technical publications.
C. It facilitates the writing task.

Corresponding benefits of job analysis to the reader include producing technical publications that are more complete, more accurate, and easier to understand and that offer vastly improved usability.

The objective of job analysis is to collect and organize information describing the operation and maintenance concept of a certain product. The technical writer is not, however, expected to perform this collection and organization of information. These tasks belong to maintenance engineering and other elements of logistic support.

The job analysis process enables the technical writer to determine the information requirements for each discrete skill level of the user. It is a critical first step in the development of the technical publication.

A Priority Sequence for Job Analysis

Job analysis also provides the technical writer with a task listing and a task analysis. *Task listings* identify each activity that is required to support the product, and *task analysis* identifies the detailed requirements that are applicable to each task. Task analysis identifies (1) the time period of performance for each task, (2) who is responsible for performing the task, (3) where the task is to be performed, (4) where the required spare and repair parts are located, and (5) what items of tools and test equipment are required for task performance.

Complete job task analysis information is frequently difficult for the technical writer to obtain. This is particularly true for the new product, which requires that the technical publication and product design be developed concurrently. In this instance, the technical writer must establish priorities based upon a time-phased sequence of information needs. This permits obtaining the most needed information at the earliest possible time. The *priority sequence* is a compromise between the needs of the technical writer and the availability of information on the developing product. Table 11.1 provides a summary of the priority of needs for the technical writer preparing a technical publication for a new product. It also includes a brief description of the activity associated with each need.

Maintenance Items

Identifying the parts upon which maintenance can be performed is the highest priority information a technical writer needs. This information, along with other engineering data such as drawings, specifications, and test reports, enables the technical writer to organize the publication and to complete a significant portion of the job analysis.

Major benefits accrue to the technical writer as a direct result of receiving this maintenance information early in the technical publication development cycle. First, since a major portion of the maintenance manual involves maintenance procedures, the technical writer can plan chapters devoted to the various levels of maintenance and to any subassemblies that require maintenance action and that merit separate chapters. The technical publication begins to take form, and the scope of writing can be assessed.

A second major benefit of receiving maintenance information early is to enable the technical writer to evaluate fault isolation requirements for each item

Table 11.1. A PRIORITY OF TECHNICAL WRITING NEEDS

Priority	Information Needed	Activity
1.	Maintenance items by category of maintenance	Organize the technical publication by chapter and section. Establish procedure requirements for each maintenance item.
2.	Test equipment and related support items	Establish general control methods.
3.	Personnel requirements	Establish who will perform the three levels of maintenance. Develop procedure outlines.
4.	Specific task identification for each item	Verify assumptions of initial task listing (priority 1). Exclude unneeded tasks.
5.	Specific task information	Determine task scheduling, frequency, time to perform, difficulty, and importance.
6.	Task data	Establish procedural steps.

requiring maintenance. For example, when parts are replaced in the assembly at the component level, fault isolation data must be included in the publication. The writer also knows that the repair technician may have to perform related activities such as alignment or adjustment of the item. Procedures relative to these activities must be included in the appropriate section of the publication.

A third benefit is that the writer is able to identify the magnitude of assembly-disassembly procedures that must be included in the publication so that the illustrated parts breakdown (IPB) and the illustrations needed to support text material can be defined.

Tools and Test and Support Equipment

The second priority—the identification of tools and test and support equipment—needs to be done early to allow the technical writer to evaluate what the reader will be capable of accomplishing at each level of maintenance activity and what equipment must be available to assist in job performance activities, since tools and test and support equipment are required for even the simplest task. These items are identified and organized by level of maintenance.

Personnel Requirements

The third priority—personnel requirements—consists of information on the personnel who will be performing the various activities and the location where this activity will take place. Location data aid the writer in organizing the publication, since separate volumes are normally prepared for each work location and for each technical specialty.

Specific Activity Data

The information provided thus far is sufficient to complete about 90 percent of the job analysis required by the technical writer. The fourth, fifth, and sixth priorities are related to the specific attributes of the tasks associated with each maintenance item at each maintenance level. This is an expansion as well as a clarification and verification of the data described in priorities one, two, and three.

Specific activity data permit verification of planned publication content by providing the answer to such questions as

A. Are all necessary tasks included within the publication and are all unnecessary tasks excluded?
B. Are the correct tools and test and support equipment referenced for each task?
C. Are the various maintenance activities assigned to the correct maintenance level and location?
D. Is the publication properly organized, and are the procedures presented in a logical sequence?
E. Does the publication include optimum procedures for the performance of each maintenance activity?

The job analysis that supports the technical writer exists on two distinct levels. The first level is the larger and more general process of determining what activity is to be performed, when it is to be performed, and where it is to be performed. The second level is the more specific process of determining detailed step-by-step procedures for the performance of each activity.

The first level has historically presented the greatest problem for the technical writer. The challenge is in locating the right information, making sure that all relevant information is included in the job performance analysis, and transferring this information into material that is appropriate to the user's skill and reading level. Once the questions endemic to this level have been successfully answered, the detailed step-by-step procedures become relatively easy to develop. The resulting publication, developed through a disciplined and coordinated job task analysis process (Figure 11.1), provides the user with what is needed to support the product.

An Overview of the Development Process

As illustrated in Figure 11.1, the publication activity is initiated when the logistics function of the firm receives a request for service. This request will normally come from within the firm as the need for a publication is triggered by a new product or system. Following input processing, the request for service is reviewed to determine the publication requirement necessary to meet customer needs. At approximately the same time, the job-task analysis activities are initiated.

Job-task analysis provides the data needed to define the scope and content of

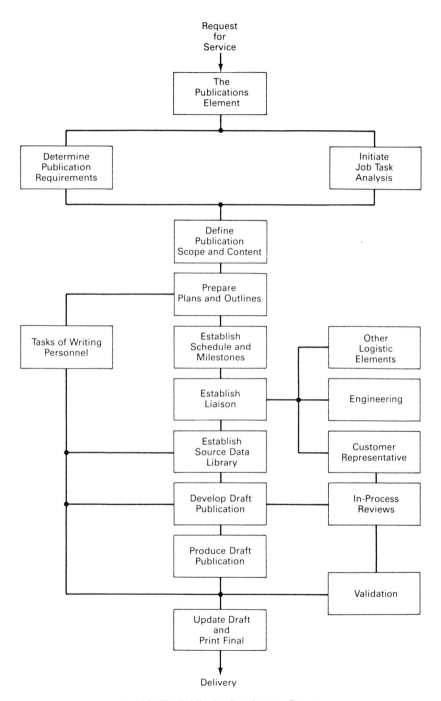

Fig. 11.1. The Publication Development Process

the publication. This permits the assignment of writing tasks to the technical writer. Concurrent with this activity, the publication schedule is developed, and milestones are established as a means of measuring progress. Communications channels between the publications element, other logistic elements, the engineering staff, and the publications representative within the customer facility are also established at this time.

A publications guidance conference between the publications element and the customer is scheduled shortly after publication planning has been finalized. This conference is to determine if the publication planning is in line with customer expectations and is an important part of the project's success.

The actual writing effort is now initiated, using the previously developed plans and outlines as a guide. Source and reference material is available from the data library, which is an organized data base incorporating the results of job-task analyses and other material relevant to the writing effort. While writing the document, in-process reviews maintain open communications between the publications element and the customer.

The writing effort culminates in the publication of a draft, which is submitted for review (validation) by appropriate individuals within the firm. The draft publication is updated as necessary, printed as a final document, and delivered to the customer.

Design of the Technical Publication

As previously stated, job-task analysis provides the raw material needed in creating a superior technical publication. Publication utility will, however, be adversely affected if adequate attention is not accorded to the design of the overall document. Visual appearance, for instance, has an almost irrational impact on the perceived usability of the technical publication. The potential reader is impressed, either favorably or unfavorably, by the visual image that is transmitted. A favorable impression is conveyed by several identifiable publication characteristics such as the use of graphics; the amount of white space on the page; and easily distinguishable chapter, section, and paragraph headings.

An equally important factor is the manner in which publication content is organized. Organization is, to a large extent, a function of the product and the type of publication. For example, the tutorial publication is intended to be read in sequence from beginning to end. The maintenance manual, on the other hand, should be organized to facilitate the search for specific maintenance and repair instructions.

An Approach to Publication Design

Technical presentations are, by their very nature, difficult to understand. This difficulty can, however, be alleviated when the completed publication has been designed and produced to promote readability. Each document should be organized to permit a cursory examination by the reader to determine what type of publication it is and what subjects are presented. A familiar format aids the process; therefore, technical publications are generally divided into the following basic units: (1) front matter, (2) chapters that make up the bulk of the document, (3) appendices, (4) an index, and (5) references.

Front Matter

In the majority of publications, front matter includes the table of contents, a preface or foreword, and the introduction. Additional items such as a list of affective pages (a listing of pages that are affected when the publication is revised) or a publication change record may be required in selected documents such as those that bear a Department of Defense (DOD) classification. These requirements are identified in the request for service when applicable.

The *table of contents* is normally the reader's introduction to the document and its contents. Since it allows the reader to approximate the potential value of the document, the well-prepared table of contents gives a favorable first impression.

A table of contents is actually a topical outline of the structure of the document. It lists the main subjects and, through heads and subheads, indicates the relationship of subtopics to each major topic. Thus the potential reader can quickly evaluate the contents. The same benefit cannot be derived from a table of contents that simply lists headings in sequential order.

The majority of technical publications are improved by a *preface* or *foreword* and an *introduction*. These sections provide the reader with a definition of the intended audience, a general description of the type of document, and the relative scope of the subject matter. This allows the reader to make a more informed decision concerning the potential value of the document in meeting projected needs. The introduction that exceeds one or two paragraphs may be included as the first chapter of the document rather than a part of the front matter.

Chapters

The various chapters constitute the bulk of the information contained in the publication. It should present a unified, logical, and coherent flow throughout the document. The reader should have a clear idea of the relationship between each chapter and section.

The first chapter should consist of either an overview of the publication or, if one was not provided as part of the front matter, an introduction. It should not, however, be vague or imprecise simply because the purpose is to present the publication content at a relatively high level. The first chapter should clearly describe the subject in general terms and indicate the relationship between successive chapters.

Each succeeding chapter should flow naturally into the subsequent chapter, thereby presenting a logical sequence to the reader. A separate chapter should be devoted to each major facet (installation, operation, organizational maintenance, and so forth) of the product that is necessitated by the purpose of the publication. References to material in other chapters must be as specific as possible to facilitate cross-referencing.

Appendices

The *appendix* can be a valuable asset to the technical publication. Many categories of data are essential to a complete understanding of the subject matter, and yet they can be out of place in the text, thereby breaking the reader's train of thought. Material such as tables of facts, data relevant to the subject, figures that amplify or enhance the material presented in the document, glossaries of terms, conversion tables, error messages, code lists, file structures, and so forth, is often appropriately included as an appendix. The exact type of material included in the appendix varies with subject matter.

Index

The *index* bears the questionable distinction of being the first item neglected when a publication runs into schedule problems. This is unfortunate since a well-constructed index is a significant improvement in a document's usefulness.

The well-constructed index should include every item of importance that is relevant to the subject of the publication. It should also indicate each place in the document that includes information about the item. Indexes are difficult to construct because importance and relevance are largely subjective and an understanding of the terms comes only with experience. The difficulty is lessened, however, when it is realized that even a sketchy index is far better than none at all.

References

References consist of listings of other published documents that may have been used as source material by the technical writer in preparing the publication. References are more commonly referred to as *bibliographies* in nontechnical books and publications.

Evaluating the Product

At the time of final production, the technical publication has been reviewed, validated, and updated. The technical content has been verified as correct, and all procedures have been tested through actual performance. The document includes the information that is needed in meeting job requirements. The product is complete. But will it be used?

Use of the publication is, to a large extent, dependent upon factors far removed from the accuracy and adequacy of the technical content contained therein. These factors, related to aesthetic appeal and ease of usage, may be evaluated by a brief review of the completed document.

The first area to review is the presentation of material in each chapter. Chapter 1 should begin with an overview or an introduction giving a general idea of the purpose of the document and what will be covered. This chapter should also lead into the remainder of the document, thereby presenting a smooth chapter-to-chapter flow. Thereafter, each chapter should begin with general information concerning chapter content. This should be followed by increasingly specific details of the topic arranged in such a manner that the reader is presented with a logical progression of thought. All the material in a chapter should be related in a logical and coherent manner, so that the totality of the chapter supports an easily understood purpose.

Graphics is another method of evaluating overall publication utility. A liberal use of graphics in support of text presentations calls attention to the material and emphasizes its importance. Graphics can lead to tremendous increases in publication effectiveness as the reader can only absorb a limited amount of uninterrupted prose.

Graphics includes illustrations and tables. Any item of equipment that can be illustrated should be, and any data that can be presented in tabular form should be. Even though the information is also presented in prose, the reader will understand it better and remember it longer if it is also presented graphically.

SUMMARY

Technical publications exist to inform. The "good" publication enables the reader to attain a high degree of performance ability, so that even the inexperienced user can, having read the publication, perform as well as the experienced technician.

Technical publications are prepared by the technical writer, whose function is to communicate to the reader by preparing a publication that is matched to the capabilities of the reader. To achieve this match, the writer must evaluate the unique capabilities of the audience the publication is intended to serve. The subject of the document, in many cases, can provide a reliable indication of these capabilities.

An evaluation of the intended audience leads to the next step, that of determining the proper scope of the publication. The author must select a subset of all information that is available, since it is not possible to incorporate all that is known about a given subject. This subset must correspond to the purpose of the planned document.

Job-task analysis is the basis for technical publications, and it provides information concerning what must be done, why it must be done, and how it must be done. This, in essence, provides the technical writer with all information needed to prepare the publication.

Technical content, in and of itself, is not enough to assure publication utility. The reader is influenced by first impressions, and the document that creates a negative impression is perceived as having little value, regardless of the quality of content. For this reason, considerable attention must be accorded the publication design. The completed product must have both consistency and visual appeal.

QUESTIONS FOR REVIEW

1. Explain why technical writers do not have to react to reader demand.
2. Why is the term *technical communicator* more appropriate than *technical writer*?
3. Define the expression "level of competence" as it relates to the technical publication.
4. Explain the difficulty encountered by the technical writer when attempting to write for readability.
5. Describe the process for defining the intended audience.
6. What is job-analysis? How does it provide the basis for a technical publication?
7. What is the purpose of the publications guidance conference?
8. Discuss the importance of visual appearance in the technical publication.
9. Discuss the importance of a well-constructed table of contents.
10. What type of material would be appropriate for inclusion in an appendix?

12

SPARES AND REPAIR PARTS

Spares and repair parts play a critical role in product support. Without spares and repair parts, the product that fails has to be discarded, with continued service dependent upon the acquisition of a new unit. Such replacement, with the exception of the inexpensive product or the one that defies repair, does not normally represent the preferred choice. The availability of spare and repair parts represents a viable alternative to this wholesale product replacement.

The logistic element called *spares and repair parts* provides for this alternative by identifying all repairable and replaceable items associated with a product. This identification process is referred to as *provisioning*. Provisioning, while normally associated with military systems, is equally applicable to the commercial firm. Provisioning, in fact, applies to any activity where products of the firm are used by the public sector, by the private sector, or by an individual.

Provisioning, by definition, is the process by which spares are identified. Spares, in turn, consist of assemblies, subassemblies, components, parts, repair kits, and raw materials that are required or are anticipated as being required as the replacement for failed items during manufacturing, operation, maintenance, repair, or overhaul. In short, spares include the total range of items and materials that may be used to restore product service.

Sparing Decision

Primary factors influencing the decision to acquire replacement items (spares) are (1) the probability that a failure will occur and (2) the consequences resulting from that failure. For example, consider the homeowner who purchases a new lawn mower. A replacement blade (spare part) is normally not acquired at the time of initial purchase. The reasoning behind this decision not to spare is the low probability of blade failure and the relatively insignificant consequences of that failure. (At most, mowing the lawn may have to be delayed for a few days.)

An additional factor relative to sparing is revealed in this example. The homeowner is not required to maintain an inventory of mower spares since the items are readily available upon demand. In effect, the firm is maintaining an inventory as a service to the consumer. The broken blade therefore impairs mower operation until it can be replaced. The ready availability of replacement items (and the low consequence of failure) renders this an effective and viable solution.

Decreased availability coupled with an increased probability of failure may, however, impact this decision in another direction. The homeowner who lives in a remote area with respect to the retail outlet and who maintains a relatively large grassy area with a lot of surface stone may decide to establish a small inventory of repair parts.

The homeowner, of course, has the alternative of discarding the defective lawn mower and replacing it with a new unit. This is one type of sparing, and it is certainly done at later stages in a product's life. For the newer unit, however, the disproportionate cost ratio provides a strong incentive toward the restoration of service through replacement of the defective part.

Ready availability does not in and of itself support a decision not to establish an inventory. Consider a consumer product such as a light bulb. This product is available at numerous outlets, and one of them is almost certain to be located near the individual consumer. Yet the average homeowner sets aside shelf space for a small inventory of light bulbs. The probability of failure and the consequences resulting from that failure (no lights) provide justification for this inventory despite the availability of replacement units.

The decision to spare is also influenced by product location and the operating environment. Consider, for example, a fleet of trucks of the type used in heavy construction. These vehicles are in the relatively benign environment of Florida. The trucks are operating eight hours per day during a normal five-day workweek, and there is no penalty for delayed completion of the project. The demands on the engine and drive train are modest. Any engine that did fail could easily be replaced through local resources, and the consequences of this failure are quite low in the absence of a penalty clause. For these reasons, it is unlikely that a spare engine would be carried in stock by the firm.

Now consider a second fleet of the same vehicles, this time operating in a Third World country. The trucks are now operating sixteen hours per day (two shifts), seven days per week, over mountainous terrain. Replacement units (spare parts) must be shipped from the United States, and there is a substantial penalty for delays in completion. Is there justification for a spare engine under these circumstances? The answer is a qualified yes. The decision to spare is partially based upon the degree of risk that the firm is willing to assume.

The preceding examples serve to illustrate some of the complexities inherent within the sparing (provisioning) process. The remainder of this chapter examines the provisioning process, the methodology used in determining what and how many of each item should be spared, and where these spares should be located.

Short History of Provisioning

Provisioning owes its existence to the U.S. government and, by extension, those firms that design, develop, and produce products for the government. Provisioning as a profession began to take shape in 1958 when select groups of government and industry personnel took steps to improve the product support posture and to establish provisioning on a more professional basis. The next several years witnessed numerous improvements and refinements in provisioning concepts and procedures.

A significant milestone in the development of a modern provisioning methodology was the publication of Department of Defense Instruction (DODI) 3232.7 in 1958. This instruction led to the establishment of comprehensive reviews, evaluations, and revisions of the provisioning procedures used throughout the military services. Provisioning data were collected, organized, and studied in initial attempts by the U.S. Army to standardize the provisioning process. AR 700-18, R-700-19, and Logistics Directive 121-750, published by the U.S. Army, represented one of the first attempts at standardization.[1] None of these early attempts could be considered a success, as each area continued to employ different provisioning concepts, techniques, and procedures. While a degree of commonality was achieved, each command in the Army retained, for the most part, independent approaches to provisioning.

A second milestone was achieved when IBM was commissioned to study and revise the provisioning process. New procedures resulting from this study were published in TM 38-717 and TM 38-715-1 in 1965.

The Department of Defense re-examined the provisioning process in the 1971–1974 time frame. The purpose of this examination was to determine the degree of standardization among the military departments and the Defense Logistics Agency (DLA). The results were not encouraging: Each military department and the DLA had established their own specific requirements, and the interchange of data was virtually impossible without an expensive and time-consuming conversion

---

[1] Department of Defense Instructions (DODIs), Army Regulations (AR), Military Standards (MIL-STD), and Technical Manuals (TMs) are publications that provide a detailed analysis and presentation of military provisioning techniques and procedures. For more advanced treatment of this subject, contact: US Army DARCOM Material Readiness Support Activity, ATTN: DRXMD-MD. Lexington, KY 40511.

process. This was equally frustrating to industry since multiple sets of provisioning documentation were required. As a result of these findings, the office of the Secretary of Defense directed the standardization of provisioning across the DOD. This direction led to the publication of three documents during 1974:

A. MIL-STD-1552, "Uniform DOD Requirements for Provisioning Technical Documentation"
B. MIL-STD-1561, "Uniform DOD Provisioning Procedures"
C. DODI 4140.42, "Determination of Initial Requirements for Secondary Item Spare and Repair Parts"

With the publication of these documents, the goal of a standardized DOD provisioning process moved a little closer to reality.

A later event having a significant impact on the provisioning process was the DOD-directed Logistic Support Analysis (LSA) program. (Logistic support analysis is discussed at length in Chapter 14.) LSA, defined as a composite of all actions taken to identify, define, analyze, quantify, and process logistics support requirements, is a key element of integrated logistic support.

Provisioning: The Government and the Firm

Provisioning as a profession is much more widely recognized in the U.S. government and with firms doing business with the government. This does not alter the fact that provisioning is also an important element in the commercial firm. The auto parts store must evaluate both the specific parts to carry (in inventory) and how many of each part should be carried. This evaluation is the provisioning process. A wrong choice in the selection of parts results in (1) items that remain on the shelf because of a lack of demand or (2) consumer demand for items that are not carried in inventory. The former condition means that financial assets are wasted on unneeded items, whereas the latter situation leads to consumer dissatisfaction, lost sales, and a possible loss of future business. A similar analogy can be drawn for every firm (or individual) utilizing an inventory of spare and repair parts.

The primary differences between government and commercial provisioning are in the methods used for financing, warehousing, and distribution. Government provisioning activities include the awarding of contracts to the firm for *provisioning documentation*. This documentation incorporates recommendations by the firm concerning the type and quantity of spares and repair parts that should be ordered. Provisioning documentation, upon approval by the government, becomes the vehicle through which the firm orders the identified spares and repair parts. The items are subsequently delivered to the firms, with the government providing for warehousing and distribution. Thus, the firm realizes a rapid return on investment, and the government assumes the burden of inventory cost.

Commercial provisioning, on the other hand, provides for the firm to determine what spares and repair parts are required. The firm must then acquire the items and warehouse or distribute them in anticipation of future demand. No return on investment is realized until the items have been sold to the consumer.

At first glance, it would appear to be financially advantageous for firms doing business with the government to encourage the sale of spares and repair costs. This, however, increases the life cycle cost, which increases the cost of products produced by the firm. The firm would be better to minimize the need for spares and repair parts, thereby reducing support costs and enhancing the overall competitive position.

Provisioning Process

The *provisioning process,* as defined in the context of integrated logistic support to the product, is the identification, documentation, procurement, and delivery of spares and repair parts. The process, illustrated in Figure 12.1, begins with a review of the product design specifications, drawings, engineering reports, and related documentation. The output of this review is the provisioning parts list (PL), which identifies each item and component that is used in manufacturing the product. Each item included in the PPL is then evaluated to determine (1) if it should be spared and (2) if spared, what the desired quantity is.

The annotated PPL represents the spares and repair parts recommendation of the firm. This list is now submitted for customer review and approval. Upon approval, the annotated list is used by the firm as the vehicle for either manufacturing or ordering the identified spares and repair parts. Upon receipt by the firm, the spares and repair parts become the property of the customer with subsequent distribution dependent upon direction received by the firm.

Early Provisioning Process

The early days of provisioning were characterized by a ritual referred to as the *provisioning conference.* The conference was established to review product data, identify items that were potential candidates for replacement or repair, and derive the recommended spares and repair parts list. Conference attendees consisted, for the most part, of maintenance and supply technicians with years of experience. Spares were selected on the basis of intuition, judgment, and personal experience. No consideration was accorded historical data, maintainability or reliability data, or a planned methodology for maintenance. Any correlation between the spares and repair parts selected as a result of this conference and what was actually required for product support was largely a matter of coincidence.

This method of provisioning, while expensive and grossly inefficient, served the purpose in an unrestricted budget environment. Unneeded items that were purchased were simply ignored, and ample dollars were available for the acquisition of needed spares that were not selected. This changed, however, with the advent of integrated logistic support, maintainability and reliability predictions, and logis-

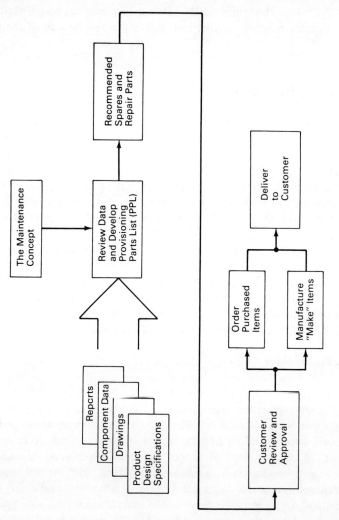

Fig. 12.1. The Provisioning Process

tic support analysis. All logistic functions, including provisioning, could now receive much needed assistance.

Logistic Support Analysis and Provisioning

Logistic support analysis (LSA) replaces the personal desires and judgment that characterized earlier provisioning conferences with a more informed decision-making process. While the conference remains a part of provisioning, judgment has been replaced with statistical data and engineering analysis.

The LSA process, illustrated in simplified form in Figure 12.2, begins with product design. This step involves the influence of integrated logistic support toward a product design that is capable of meeting logistic support requirements. Activities during this phase include initial logistic and repair-level analysis. Any steps that may be taken to increase product reliability and maintainability must be implemented during this design phase.

Finalization of the design leads to the second phase of logistic support analysis, which incorporates detailed maintenance engineering, including reliability and maintainability analysis, into the maintenance plan. LSA data, which quantify the total product support requirements, are stored as a logistic support analysis record (LSAR). The LSAR is normally stored in a computer data base, although the data base may consist of nothing more than printed forms stored in a file cabinet. Regardless of data base form, the LSAR contains all information (item population, reliability factors failure rates, repair levels, and so forth) needed to derive the recommended spares and repair parts list.

ILS involvement during the early stages of product development provides an excellent opportunity for the design of an optimum logistic support package. This rarely happens, since few, if any, organizations include ILS as a viable organizational function.

A Provisioning Example

The provisioning process requires a knowledge of (1) the population of the item being spared, (2) the operating scenario, and (3) the projected failure rate. The use of these data in the provisioning process may be illustrated by considering a common product such as the ordinary incandescent light bulb used in the typical household. The light bulb cannot be repaired; therefore, it is spared by replacement of the entire unit. For the purposes of this illustration, the following are assumed:

A. The household includes a total of twenty-two light bulbs.
B. Each light bulb is in use an average of 4 hours each day, 7 days a week.
C. The mean time between failures is 800 hours.

Using these assumptions, each light bulb is on for 28 hours per week (4 hours per day times 7 days) or 1,456 hours per year. The cumulative time for all

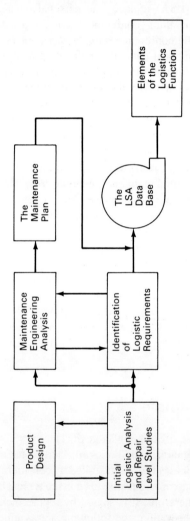

Fig. 12.2. A Simplified LSA

22 light bulbs is, therefore, 32,032 hours (1,456 hours per light bulb multiplied by 22).

The mean-time-between-failure for a given light bulb is 800 hours, which is equal to a failure rate of 1.25 failures for each 1,000 hours of operation. The total operating time per year (32,032 hours) multiplied by the failure rate of 1.25 failures in each 1,000 hours of operation yields 40 failures that may be expected to occur throughout the year. A one-year supply of bulbs would then require the acquisition of 40 spare units. This same number could have been derived by dividing total operating hours (32,032) by the mean-time-between-failure (800 hours).

The fallacy of this logic is the assumption that no light bulb will fail in less than 800 hours of operation. Mean-time-between-failure figures represent statistical averages that have been derived from a large population. The operational life expectancy of a single unit, however, may range from a few moments to some period of time exceeding the mean-time-between-failure average. The acquisition of only 40 spare units does not allow for any failures before 800 hours of operation (that is, there is no safety stock). Of course, it is equally possible that the operating life could exceed the mean-time-between-failure average and that 40 spare units will exceed need. This, while it represents an unnecessary expenditure of assets, is a less serious problem than too few of a critical item.

The ideal situation is one in which replacement and spare items match failures, and there is neither a shortage nor a surplus of spares. Such accuracy in the prediction of future events is unlikely; therefore, some amount of safety stock is normally incorporated in the recommended spares quantity. The amount of safety stock is a function of both financial constraints and the degree of risk that the firm is willing to accept. Risk is evaluated through statistical analysis and expressed as a percentage certainty. For example, a quantity of safety stock units representing two standard deviations provides approximately 95 percent certainty that there will not be a stock-out.

Sparing Determinants

Logistic support analysis provides the data necessary to accomplish the provisioning task in a logical and systematic manner. Additionally, the reliance on engineering analysis provides results that are inherently more accurate than decisions based on human judgment. Two key steps in the provisioning process are enhanced through the use of LSA data: (1) the determination of source, maintenance, and recoverability (SMR) codes and (2) the determination of required sparing quantities.

SMR Coding

SMR codes are 5- or 6-letter codes, developed by the U.S. government, in which each letter and its position describe specific attributes of the part being addressed. The codes apply to any firm doing business with the U.S. government where provisioning is a part of the contract. The code lists identified in Figures 12.4 and 12.5 are standard throughout industry. Such codes define the acquisition, use, and method of disposal of each item identified in the provisioning parts list. Figure 12.3 illustrates their format. Source codes (positions 1 and 2) define the process for acquiring the item, that is, make, buy, assemble, salvage, and so forth. Maintenance codes are incorporated in positions 3 and 4. Position 3 indicates the lowest maintenance level (organizational, intermediate, or depot) that is permitted to remove or replace the item. Position 4 identifies the level of repair, and position 5 (recoverability) identifies the lowest maintenance level that is permitted to dispose of the item. Position 6 is optional. This position is used by the U.S. Air Force to define the expendability, repairability, and recoverability code (ERRC). The ERRC is used to identify the cost category and expendability of items. Figures 12.4 and 12.5 illustrate the codes used in Joint Military Services SMR coding. The primary differences are the use of position 6 by the U.S. Air Force and additional codes used by the U.S. Navy to differentiate between land and sea capabilities.

The provisioning process requires the development of an SMR code for each item (part) that makes up the product. The modern military system includes hundreds of thousands of individual parts. SMR coding is thus a formidable task. SMR codes represent the link between the LSA and the user by identifying the approved maintenance concept, which is derived through the LSA process.

A Sample SMR Code

Let us consider the code designation for the light bulbs used in the provisioning example. The applicable SMR code is illustrated in Figure 12.6 and illustrates the use of SMR codes in defining the maintenance concept. The code defines (1) how to obtain a replacement part, (2) the level of maintenance authorized to replace the part, (3) the level of maintenance for repair of the part, (4) whether the part can be repaired, and (5) the maintenance level that is required to condemn the part. As

SOURCE CODES	MAINTENANCE CODES		RECOVERABILITY CODES	
Positions 1 and 2	Position 3 (Use)	Position 4 (Repair)	Position 5	Position 6
How is the support item to be acquired?	What is the lowest maintenance level authorized to remove and replace the item?	Is the item a repair or discard? If repair, what is the lowest level of maintenance that is authorized to perform the repair?	What level is authorized to determine disposition of the item?	Reserved for service option

Fig. 12.3. SMR Code Format

SOURCE			MAINTAINABILITY			RECOVERABILITY		AF ERRC CODE	
POSITION 1		POSITION 2		POSITION 3		POSITION 4		POSITION 5		POSITION 6
P	PROCURABLE	A	STOCKED	O	REPLACE AT O-LEVEL	Z	NO REPAIR	Z	NO REPAIR: CONDEMN AT ANY LEVEL	N	NOT RECOVERABLE CONDEMN AT ANY LEVEL
		B	INSURANCE								
		C	AGE								
K	PART OF A REPAIR KIT	F	I-LEVEL KIT			B	RECONDITION	O	REPAIRABLE: CONDEMN AT O-LEVEL	P	RECOVERABLE: UNIT COST LESS THAN $100
		D	DEPOT KIT								
		B	BOTH KITS								
M	MANUFACTURE	O	AT O-LEVEL	F	REPLACE AT I-LEVEL	O	REPAIR AT O-LEVEL	F	REPAIRABLE: CONDEMN AT I-LEVEL	X	RECOVERABLE: UNIT COST MORE THAN $100
		F	AT I-LEVEL								
		D	AT DEPOT								
A	ASSEMBLE	O	AT O-LEVEL	D	REPLACE AT DEPOT	F	REPAIR AT I-LEVEL	D	REPAIRABLE: CONDEMN AT DEPOT	T	RECOVERABLE: CONDEMN AT DEPOT
		F	AT I-LEVEL								
		O	AT DEPOT			D	REPAIR AT I-LEVEL OVERHAUL AT DEPOT	L	REPAIRABLE: CONDEMN AT DEPOT ONLY	C	RECOVERABLE: SCANS
Y	NON STOCKED	A	NEXT HIGHER ASSEMBLY								
						L	REPAIR AT DEPOT	A	SPECIAL HANDLING REQUIRED	S	NOT EXPENDABLE
		B	SALVAGE								
		C	SEE DRAWINGS							U	INTERMEDIATE: NOT EXPENDABLE

Fig. 12.4. Air Force Use of Joint Military Services Uniform SMR Coding

SOURCE			MAINTAINABILITY			RECOVERABILITY	
POSITION 1	POSITION 2		POSITION 3		POSITION 4	POSITION 5	POSITION 6
P PROCUREMENT	A	STOCKED	O	REPLACE AT O-LEVEL	Z NO REPAIR	Z NO REPAIR: CONDEMN AT ANY LEVEL	NOT USED
	B	INSURANCE					
	C	LIMITED LIFE					
	D	INITIAL ISSUE				O REPAIRABLE: CONDEMN AT O-LEVEL	
	E	GSE/STOCKED	F	REPLACE AT I-LEVEL ON-SHIP	B RECONDITIONED		
	F	GSE/NON-STOCKED					
	G	SUSTAINED SUPPORT			O REPAIR AT O-LEVEL	F REPAIRABLE: CONDEMN AT I-LEVEL ON-SHIP	
K PART OF A REPAIR KIT	F	O/I LEVEL KIT	H	REPLACE AT I-LEVEL: ON-LAND	F REPAIR AT O-LEVEL: ON-SHIP	H REPAIRABLE: CONDEMN AT I-LEVEL ON-LAND	
	D	DEPOT KIT					
	B	BOTH KITS			H REPAIR AT I-LEVEL: ON-LAND		
M MANUFACTURE	O	O-LEVEL	G	REPLACE AT I-LEVEL: BOTH	G REPAIR AT I-LEVEL: BOTH	G REPAIRABLE: CONDEMN AT I-LEVEL: BOTH	
	F	I-LEVEL -- ON SHIP					
A ASSEMBLE	H	I-LEVEL: ON-LAND	D	REPLACE AT DEPOT	D REPAIR AT DEPOT	D REPAIRABLE: CONDEMN AT DEPOT	
	G	I-LEVEL: BOTH					
	D	DEPOT	L	REPLACE AT SPECIALIZED REPAIR FACILITY	L REPAIR AT SPECIALIZED REPAIR FACILITY	L REPAIRABLE: CONDEMN AT SPECIALIZED REPAIR FACILITY	
X MISCELLANEOUS	A	NEXT HIGHER ASSEMBLY	Z	DO NOT REMOVE OR REPLACE		A REQUIRES SPECIAL HANDLING	
	B	SALVAGE					
	C	SEE DRAWINGS					

Fig. 12.5. Navy Use of Joint Military Services Uniform SMR Coding

repaired, and (5) the maintenance level that is authorized to condemn the part. As is apparent from these examples, SMR coding is a critical element of the provisioning process. It should only be performed by experienced provisioning professionals.

Spares Quantity Determinants

The determination that a given item should be spared is followed by a determination of the required spares quantity. Once again, the LSA provides all the data required to arrive at this number.

The basis for determining the quantity of spares is the reliability data expressed in the LSA. These data, however, require modification before they can be used in spares computation, because reliability figures reflect primary failures only. It does not reflect failures that are caused by other parts in the system or by external conditions such as human error. These types of failure create a demand on the spares inventory and must also be considered when determining spares quantities. This is accomplished by developing a *mean-time-between-demand* (MTBD) figure.

The first step in developing an MTBD figure is to reduce the mean time between failures to a figure known as the *mean time between unscheduled maintenance actions* (MTBUMA). This is accomplished by the incorporation of secondary failures. For example, the previous light bulb example assumed a mean-time-between-failures figure of 800 hours (a failure rate of 1.25 failures for each 1,000 hours of operation). If it is further assumed that one light bulb is dropped (broken) for every four bulbs that are replaced, the mean time between failures due to breakage (secondary failures) is 3,200 hours or a failure rate of 0.3125 failures for each 1,000 hours of operation. The combined (primary failures plus secondary failures) mean time between failures therefore becomes 640 hours. This process is illustrated in Figure 12.7.

An MTBUMA of 640 hours equates to a failure rate of 1.5625 failures for each 1,000 hours of operation. Substitution of this figure yields a spares requirement of approximately 50 units (32,032 total operating hours times the failure rate of 1.5635 failures for each 1,000 hours of operation). This is an increase of 10 units over the 40 determined as necessary when the primary failure rate was used. This number also must be increased by the desired quantity of safety stock.

Fig. 12.6. An Example of SMR Coding

$$\text{MTBUMA} = \frac{\text{Primary MTBF} \times \text{Secondary MTBF}}{\text{Primary MTBF} + \text{Secondary MTBF}}$$

$$= \frac{(800)(3{,}200)}{800 + 3{,}200}$$

$$= \frac{2{,}560{,}000}{4{,}000}$$

$$= 640 \text{ hours}$$

Fig. 12.7. MTBUMA Calculations

Secondary failure rates are derived through judgment and the analysis of historical data on similar products or equipment.

Budget Constraints

In spares computations we have thus far ignored the potential impact of budget constraints upon the provisioning process. Spares quantities have been developed in accordance with the demand for support resources at each level of maintenance. In real-world situations, however, the financial resources available for the acquisition of spares may be less than that required. Computed spares costs may equal $25 million, yet only $22 million may be available. Obviously a method is needed to use the budget limitations to optimum value.

Spares can only be purchased up to the limits of the budget. Therefore, the obvious choice is to select a prioritized subset of the total spares complement. Priority of purchase is based on reliability: The parts that are most likely to fail are purchased before those that are less likely to fail. This purchase by ever-decreasing probability of failure continues until the spares budget has been expended.

It must be pointed out, however, that a decrease in the spares allotment is reflected in increased out-of-service time or extended repair cycles or both. This must be accepted in the real world with its finite financial resources.

Postproduction Support

Although products are in production during a portion of their life cycle, contingency planning must recognize that the product, or selected assemblies within the product, may be out of production at some time in the future. This is particularly true in the military services where older systems undergo modernization that effectively extends the product life cycle. The *out-of-production product* is a product whose demand decreases to the point where a production capability cannot be maintained. Under these conditions, both human and physical resources are applied to alternate tasks.

There is a significant difference between the support of products that are in production and those that are out of production. Resolution of these differences requires changes in the provisioning process if effective logistic support is to be continued. Postproduction planning, if initiated prior to cessation of production, alleviates many of the problems endemic to this phase of the product life cycle.

Postproduction Spares

The new product and the existing product that is still being produced are relatively easy to support. Spares are normally manufactured concurrently with the product. Start-up problems are solved in a relatively short time, and the spares assets soon begin to flow.

The benefits that accrue from the support of products in production are numerous. Set-up costs for the manufacture of spare parts are minimized, and material may be purchased in larger quantities at a lower cost. These large purchases lead to buying power, which provides leverage in the competition with other firms for priority shipments. Volume, current technology, asset availability, human resources, and dedication are characteristics of "in-production" spares.

The situation is dramatically different when products are no longer in production. The benefits available during the in-production phase are gone. Volume is low, since postproduction spares orders by the government, for example, are for less than ten units per contract line item in the majority of instances. This low volume negates productivity and does not allow the firm a reasonable margin of error in the manufacturing and test process. The inevitable result is escalating costs and greatly increased delivery schedules. Small quantity purchase requests face the same problems of high costs and long schedules when orders are placed with the firm for spares and repair parts that are a part of inventory.

Modernization programs that increase the operational life cycle of the "older" product add to the problems of postproduction support. The old technology associated with these systems creates problems in maintaining test equipment and manufacturing resources. Additionally, it may be difficult to locate people who are skilled in these technologies.

A mixture of orders for high-volume in-production spares and low-volume out-of-production spares has a very predictable result. The out-of-production spares are assigned a much lower priority by the firm. It is no coincidence, then, that over 75 percent of delinquent spares orders are for out-of-production items.

Postproduction Planning

The problems of postproduction support may be alleviated by recognizing the problem and planning to assure the continuity of product support throughout its life cycle. The vehicle for meeting the demands of postproduction support is the post production continuity plan.

Postproduction support continuity may be considered as a management activity dedicated to solving the spares and repair parts needs following production. The objective is to achieve a contractually binding agreement between the firm, suppliers to the firm, and the customer to provide a vehicle for lowering support costs through a reduction in the cost of spares. The plan should also address logistic support alternatives that have potential for reducing support costs.

SUMMARY

Spares and repair parts include all items associated with a product that may be repaired or replaced. The identification of these items is a major logistic activity referred to as provisioning. Provisioning acts to identify both the items that are to be spared and the quantity of the item that should be spared.

Provisioning as a profession is largely limited to the U.S. government and firms that do business with the government. Provisioning also plays a significant role throughout the entire private sector.

Two primary factors that influence the provisioning process are the probability of failure and the consequences that may result from that failure. Additional factors include the availability of spares, the operating environment, and the cost of spares relative to product cost.

Provisioning was literally created by the U.S. government in an attempt to standardize the sparing process. The process of standardization was initiated in 1958 with the publication of DODI 3232.7. Several improvements and refinements were implemented in succeeding years, culminating in the DOD-directed Logistic Support Analysis Program. The LSA Program replaced the previous decision-making process based on human judgment with one using engineering analysis.

Two key steps in the provisioning process are the development of source, maintenance, and recoverability (SMR) codes and the determination of spares quantities. SMR codes are assigned to each item that is included as a part of the product or system. They provide information concerning how the individual item is acquired, the lowest maintenance level that is authorized to remove or replace and repair the item, and the lowest maintenance level authorized to initiate disposal of the item. The assignment of SMR codes is a critical task that should be performed by an experienced provisioning professional. Data needed for the determination of SMR codes and for a determination of sparing quantities are extracted from LSA.

Spares quantity determination is based upon reliability data. Reliability data must be modified, however, since the reliability measure (mean time between failures) is related to primary failures only. Spares inventory demand is also influenced by secondary failures or those failures caused by exterior (to the item) events. Secondary failure rates are derived through the analysis of historical data. Primary and secondary failures are then combined to derive the mean time between unscheduled maintenance actions. This measurement is a more realistic determination of spares quantities. This approach does not include an allowance for any safety stock units, however. These, if desired, must be added to the spares quantity that is purchased.

Problems faced by provisioning include the real-world constraint of budgets and the issue of postproduction support. Budgets may require reducing the spares purchase below the quantity determined as necessary in providing effective support. In this instance, spares should be acquired on the basis of priority until available funds are expended. The reduced complement of spares will, however, result in a degraded support capability.

Postproduction support is concerned with the acquisition of spares and repair parts resources following production of the product. This phase is characterized by a lower spares demand, which results in greatly increased costs and extended delivery schedules.

QUESTIONS FOR REVIEW

1. Describe the provisioning process.
2. What are the major factors that influence the decision to acquire spares and repair parts?
3. What is the major contribution to provisioning by logistic support analysis?
4. Describe the differences between government and commercial provisioning.
5. What are SMR codes?
6. How is safety stock accounted for in the provisioning process?
7. What are the differences between primary and secondary failures?
8. Develop and explain an SMR code for a familiar product or item.
9. What is postproduction support?

13

TEST AND SUPPORT EQUIPMENT

Common questions that are frequently asked in relation to test and support equipment are "What is it?" "Why is it needed?" and "What does it do?" In answer to the first question, test and support equipment may be defined as the aggregate of all tools, monitoring equipment, diagnostic and checkout equipment, calibration devices, servicing and handling equipment and maintenance aids such as work benches that are used in support of the product. Test and support equipment is used throughout the total range of product support including all scheduled and unscheduled maintenance activities. The requirements for test and support equipment are derived through the logistic support analysis (LSA) process, published in the maintenance plan and differ for each level of maintenance.

Product adjustment, service, and repair constitute the principal activities requiring the use of test and support equipment. Test and support equipment, if it is to facilitate the support activities, must be designed or selected to complement the product, the operating environment, and the limitations of product support personnel. It is not unusual for the test or support equipment to be more complex than the product being supported. Test and support equipment quantities are a function of (1) the product population being supported, (2) product reliability and maintainability which incorporates the expected number of maintenance actions and the time to effect repair, (3) the maintenance concept and (4) the number of maintenance locations at each level of maintenance. All of these data are derived from the LSA.

Test and Support Equipment Categories

Test and support equipment ranges from simple hand tools and gauges to highly sophisticated and complex test, measurement, and diagnostic equipment (TMDE). These items require support from each logistic element within the logistics function in the same manner as the product itself.

Test and support equipment is unique in that it does not clearly fall within the category of spares and repair parts or any of the other product support elements. It is certainly required for maintenance of the product, yet various subsets of the test and support equipment complement are required if the product is to be moved, if product training is to be provided, and for any modifications or enhancements to the product. In short, test and support equipment is required, for one reason or another, throughout all stages of the product life cycle.

Test and support equipment can be divided into two discrete categories, as follows:

A. Common (or general) test and support equipment. These are items of equipment that are currently in general use. They are not related to a specific product and may be used with equal facility on a variety of products.
B. Special (or peculiar) test and support equipment. This category includes items that are not in general use or have been specifically designed to provide support to a particular product. While this special equipment may be adapted for use on a range of products, it is normally associated with the product responsible for its design.

Both categories of test and support equipment are identified and defined through programs designed to assure that the right types and quantities of equipment are at the right place when needed. Product maintenance, for example, requires this equipment in order to perform scheduled and unscheduled maintenance actions.

Levels of Maintenance

Test and support equipment, which can be anything from a piece of wire to a complex test system incorporating a computer as a test and diagnostic device, is selected or designed to support a level of maintenance activity. The equipment can be provided for support of one maintenance level or any combination of the three maintenance levels (organizational, intermediate, or depot).

Organizational Maintenance

Organizational maintenance is normally performed at the operating location by the personnel who use the product. This level of maintenance is frequently performed by operating personnel rather than maintenance technicians. For this reason, it is sometimes referred to as *user maintenance* or the *user level*. Maintenance activities at this level consist, for the most part, of periodic checks of product performance, visual inspections, cleaning, limited servicing of the product, and external adjustments. Product repair at the organizational level is limited to malfunctions that can be corrected through the removal and replacement of readily accessible assemblies or components. The objective of this maintenance level is to quickly restore the product to operational service.

Organizational maintenance personnel (whether operators or maintenance technicians) generally do not repair the defective unit. This repair is properly a function of higher levels of maintenance (intermediate or depot).

Intermediate Maintenance

Intermediate maintenance includes the repair of defective units and assemblies that have been removed at the organizational level. The defective item is repaired through the identification, isolation, and replacement of major assemblies and piece parts. Intermediate maintenance personnel are generally more skilled and better equipped (with test and support equipment) than those at the organizational maintenance level. They are, therefore, responsible for the performance of more detailed maintenance activities. Repair actions exceeding this level of capability are forwarded to the depot level.

Depot Maintenance

Maintenance at this level constitutes the highest type of maintenance support. It is expressly dedicated to those tasks that are above and beyond the capabilities (skills, test and support equipment, facilities, and parts) of the lower two levels. This level includes the complete overhaul and rebuilding of defective units as well as the performance of very complex maintenance actions.

The three levels of maintenance are normally associated with specific locations, each with an inventory of spare and repair parts appropriate to the maintenance level. Organizational maintenance, for example, is performed at the operating location, whereas intermediate maintenance is performed at a facility near the operating location and depot maintenance is performed at remote locations. This is generally true due to the allocation of test and support equipment that, to a large extent, dictates the maintenance capability. Levels of maintenance, however, constitute an activity level rather than a location. Replacing a light bulb is an organizational maintenance activity, whether it is replaced at the operating location, an intermediate maintenance facility, or a depot.

Each maintenance level also incorporates elements of both preventive maintenance and corrective maintenance. *Preventive maintenance* refers to all scheduled maintenance actions that are performed to retain the product in a specified condition. Operating condition is intentionally omitted from this definition, since preventive maintenance may, for example, apply to the product that is retained in storage for use at a later time. Preventive maintenance activities include inspections, monitoring of conditions, item replacement (for example, an oil filter), and service routines such as lubrication or fueling. Since these activities occur on a regular basis, preventive maintenance is also referred to as *scheduled maintenance*.

Corrective maintenance, on the other hand, is by nature unscheduled and initiated upon a product failure. The objective of corrective maintenance is the rapid restoration of service. It includes the maintenance actions of failure identification, isolation, and repair. Repair actions, which may include disassembly, removal, replacement, or repair and reassembly, are followed by checkout and condition status verification. It should be apparent that corrective (unscheduled) maintenance actions may occur as a result of a suspected failure, even though additional investigation reveals that there was no actual failure.

Requirements for Test and Support Equipment

The product, regardless of the quality of design, is going to fail. Attention paid to reliability helps to increase the time between failures, and time spent on maintainability helps to decrease the time required to effect repair; however, failure will eventually occur. Failures are of various types, and many of them are induced by events only indirectly related to the product. The types of product failures that may be anticipated include the following:

A. Primary failures. These are failures based upon physical characteristics of the product and the environment in which it operates. Primary failures may be projected through reliability data and are assumed to occur at a constant failure rate.

B. Manufacturing defects. The new product is often characterized by a high failure rate during the initial operating period. This "infant mortality" continues until the product has been in operation for a period of time and the stable failure rate is achieved. Initial support considerations must incorporate a reserve capability that is adequate for this period.

C. Operator-induced failures. Product failures resulting from operator error are a distinct possibility and must be considered when planning support needs. These failures may be reduced significantly through the use of human factors engineering during product design phases.

D. Maintenance-induced failures. Failures resulting from human error during maintenance activities represent another possibility. These are usually caused by incorrect maintenance procedures or practices, improper test and support equipment, loss of parts, or failure to reinstall all parts prior to assembly of the defective item.

E. Dependent failures. Dependent (or serial) failures are secondary failures that occur as a result of a primary failure (the failure of one item leads to the failure of one or more additional items). This type of failure may also be operator- or maintenance-induced.

F. Wear-out failures. Wear-induced failures are related to the continuous use of components. These normally occur following extended periods of operation or near the end of the product's useful life. They may be reduced, with a concurrent increase in useful product life, through the rigid application of quality standards during the production process.
G. Damage-induced failures. This category of failures includes those resulting from damage during handling (bumping, dropping, and so forth). The probability of these failures is particularly high when the product is being transported.

Regardless of the type or cause of failure, the net result is that a product is out of service. The customer, who has committed substantial resources to the product and the necessary service that it provides, cannot afford an inoperative system.

Economic justification requires that the product remain in service over the major portion of its life cycle. Obtaining this maximum operating time requires minimizing the amount of time the product is out of service for maintenance. The product is of limited value if frequent failures require lengthy periods of time for the location, diagnosis, and repair of defective items. Test and support equipment provides an inherent ability to monitor, control, test, measure, evaluate, repair, and calibrate the product, thus providing the resources needed for a rapid return to service.

Maintenance requirements are identified through a logistic support analysis program and presented in the maintenance plan. The *maintenance plan* is a document presenting a detailed plan for maintenance support. It specifies the maintenance resources and methods and procedures required for support throughout the product life cycle.

The identification of a maintenance requirement does not, however, lead to an automatic assumption that test and support equipment is required. The identification of probable failure modes and maintenance needs should also consider alternative approaches. For example, the product may be designed with a built-in test (BIT) capability or built-in test equipment (BITE). This approach may provide a self-contained capability for fault isolation of the defective assembly at the organizational maintenance level. This may effectively eliminate intermediate-level maintenance. The advantages of this self-contained capability must be weighed against the added cost and complexity of the product before a decision is reached to proceed with either BIT, BITE, or test and support equipment.

A Test and Support Equipment Program

The test and support equipment program has the objective of eliminating the need for developing special items of test and support equipment. While the total elimination of special items may not be possible in all cases, even partial successes have the potential of greatly reducing product cost. A secondary objective is to minimize the quantity and type of test and support equipment that is required for maintenance activities.

The objective of eliminating the need for special items can only be realized by a continuing review during early product design phases. Only at this stage of product development can design recommendations be incorporated to reduce or eliminate the need for special test and support equipment. A simple method of meeting this objective is the liberal inclusion of test or monitor points. This permits stage-to-stage evaluation of product performance, thereby eliminating the need for elaborate "system-level" support items.

The second objective is easily attained through monitoring maintenance planning activities. This prevents individual engineers from selecting the same items capable of performing the same function. In the absence of this monitoring, each engineer has a tendency to select the test and support items that appeal to his or her personal bias.

Selection Process

The selection process for test and support equipment is concurrent with product design and development. It begins with the customer request for a product that is capable of meeting an assigned objective and culminates with final product design. The sequence of events contributing to the selection process are summarized in Figure 13.1. The first step in this process involves an evaluation of the customer request and the product objective or operational requirements. This analysis provides an indication of basic functions and parameters that must be considered and insight into the operating environment. All these factors may be used in the generation of probable test and support equipment requirements.

The next step is to evaluate the evolving design to ensure that maintainability and maintenance needs are accorded proper emphasis throughout the design phase. This is followed by a determination of preventive maintenance requirements and a detailed analysis of probable failure modes. This latter process identifies the requirements for corrective maintenance activities. Questions to be considered throughout the selection process include

A. Is there a definite need for each item of test and support equipment? Is there any possibility of eliminating selected items through design changes? Can any of the maintenance activities be performed using simpler items of test or support equipment?
B. What are the environmental conditions? Is the product operated in an enclosed environment or outside? Is the product mobile and, if so, how frequently are moves expected? Can the associated movement and handling equipment be reduced in scope or complexity?
C. What are the parameters to be measured by test equipment? What accuracies and tolerances are required? Is there existing test equipment at the same

location that is capable of performing the required task? Would it be economically feasible to refer the activity to a higher level of maintenance?

D. Can the support task be performed by existing off-the-shelf test and support items as opposed to the development of specialized equipments?

A test and support equipment program that properly addresses the questions listed above minimizes the equipment required for product support. This does not eliminate all items of special test and support equipment. Advances in technology always create situations wherein existing items of test and support equipment are not capable of performing the product support task. This, however, is the exception rather than the rule.

Requirements for specialized items that survive the analysis inherent within these questions represent a legitimate need for equipment that is out of the ordinary. Steps must now be taken for the development of this equipment.

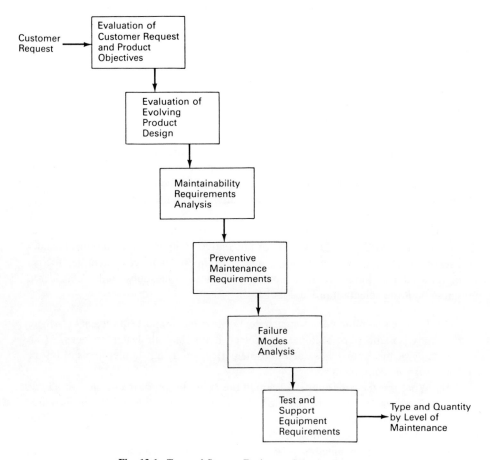

Fig. 13.1. Test and Support Equipment Selection Process

Quantity Determination

Test and support equipment identification must be followed by quantity determinations. The objective of this step is to identify and minimize the items that are required at each level of maintenance. In completing this step, the final item list is divided into test and support equipment requirements at each of the three maintenance levels. The resulting list should then be compared with the item complement currently existing at the three locations where this maintenance is to be performed. Any duplication may reduce the total required quantity.

The test and support equipment list is now reviewed with respect to the interrelationship among the levels of maintenance. Most test and support equipment can be used at more than one level of maintenance, with some items being used at all three levels. Analysis may reveal that some maintenance activities should be performed at another level. This has the potential of eliminating some items at one or more maintenance levels, thereby reducing overall support costs.

Logistic Requirements

In the past, test and support equipment has been largely ignored when performing logistic evaluations addressing support to the product. It must be remembered that these items require the same degree of support as the product or systems. A product failure cannot be corrected if the needed test and support equipment is either out of service or was not provided in the first place. Logistic requirements for test and support equipment must be planned for and implemented with the same consideration that is accorded the product.

Test and support equipment requires the same inventory of spares and repair parts that is maintained in support of the product. The majority of items also require technical publications, trained personnel, and maintenance at one or more of the three levels. Operation and maintenance procedures for test and support equipment must be analyzed to define any activity that may require specialized test and support equipment. The analysis must address what maintenance activities are required, who is to perform them, where are they to be performed, the requirements for inventory, and the required personnel skill levels.

SUMMARY

Test and support equipment has historically been relegated to a degree of logistic support that is much less than that accorded the product. It was usually forgotten in the storeroom or on a shelf until needed. Then it was expected to work and work properly. Test and support equipment exists to maintain operational service of the product, but who maintains the test and support equipment? This question usually went unanswered.

Test and support equipment incorporates the resources necessary to perform product adjustment, service, and repair. In this capacity, it must be selected or designed to complement the product. The selection of test and support equipment should follow product development to assure this compatibility. An initial review of the product specifications reveals the parameters that must be measured and the operational scenarios related to product use. This information is sufficient to begin preliminary analysis of the test and support equipment inventory.

The involvement of logistics throughout product design and development is necessary to assure that proper emphasis is placed upon maintenance and maintainability characteristics. It is only at this stage of the product life cycle that design changes may be introduced to decrease the need for test and support equipment and the need for specialized equipment items or to eliminate the need for specialized support items altogether.

Test and support equipment requirements are derived from logistic support analysis data in the same manner as data are extracted for support of the product. These requirements are a function of the product population being supported, the expected failure rate, the time to effect repair, and the number of planned maintenance facilities.

The listing of test and support equipment should be complete at the time product design has been finalized. Activities beyond this time involve the allocation of items by maintenance level and reviews to assure that duplicate items have been eliminated.

Logistic support is required for test and support equipment just as it is for the product that is the primary objective of support. This equipment requires the support of all logistics elements if it is to fulfill the objective of minimizing product downtime.

QUESTIONS FOR REVIEW

1. Define the two categories of test and support equipment.
2. What data are considered when determining the quantities of test and support equipment that will be required to support a product?
3. Describe a method by which the requirement for specialized items of equipment may be eliminated.
4. What are the primary activities supported by test and support equipment?
5. Are items of test and support equipment designed for specific levels of maintenance? Explain your answer.
6. What are some of the steps that may be taken to reduce the quantities of test and support equipment?
7. Define the three levels of maintenance using an ordinary product such as the automobile.
8. Why is test and support equipment needed?
9. What is the economic justification for test and support equipment?
10. What are the logistic requirements required by test and support equipment?

14

LOGISTIC SUPPORT ANALYSIS

Logistic support analysis (LSA) is an iterative analytical process that identifies the logistic requirements necessary for support of the new system. LSA incorporates the use of quantitative methods to assist (1) in the definition of logistic standards which should be reflected in product design, (2) in the evaluation of potential design alternatives, (3) in the identification of needed support factors that are provided by the various logistic elements, and (4) in an evaluation of product support capabilities throughout the product life cycle.

The LSA process is almost exclusively the province of the U.S. government and firms that do business with the government. It is a relatively expensive program that requires the expenditure of today's financial resources to achieve reductions in the future costs throughout the product life cycle. The cost of LSA may, however, be partially or even totally offset through data maintained in the logistic support analysis record (LSAR), which offers significant reductions in the cost of other logistic elements (technical publications, training, provisioning, and so forth). The costs associated with LSA will pay much greater dividends in the future through major reductions in life cycle costs and measurable improvements in the quality of logistic support.

LSA is the single analytical logistics effort within the product design and development process. LSA methodologies provide the tools necessary to identify, define, analyze, quantify, and process all integrated logistic support (ILS) requirements. LSA data form the basis for product design versus product support trade-off studies, the establishment of logistic support requirements, and the development of a logistic support package that incorporates the resources necessary for effective life cycle support. A primary objective of LSA is the development of logistic resources that optimize product support at an affordable cost.

An effective LSA program integrates the logistic elements into a logistic support system that is capable of meeting all the demands involving the product. The successful implementation of such a program demands the establishment of a free and continuing information exchange between the LSA staff and the engineering disciplines. This interchange is essential to the achievement of an optimum balance between the product, its operational capability, logistic requirements, and cost. The keys to a successful LSA program are (1) planning to identify the required actions, (2) scheduling to provide the proper timing for all required actions, and (3) executing each required action through timely management intervention.

Objectives of LSA

The primary objective of LSA has been defined as the development of logistic resources that optimize product support at an affordable cost. This objective can be successfully completed by taking four steps that support the overall LSA objective. These steps are as follows:

- A. The establishment of a logistic influence on product design. The LSA program must exert an influence on product design and development that is based upon logistic considerations. Initial analysis efforts are directed toward an evaluation of the effects of alternative approaches. Initial logistic requirement or goals are established based upon known constraints and logistic risks. Subsequent analysis is directed toward the refinement of initial assumptions to provide greater ease of maintenance, potential reductions in support requirements, and improvements in supportability.
- B. The identification of logistic issues and requirements. The effective LSA program identifies both quantitative and qualitative logistic support issues and requirements. Comprehensive analytical techniques are used to translate alternative concepts and operating parameters into support concepts, support costs, and supportability.
- C. The effective communication of logistic requirements. The LSA program communicates logistic requirements through the integration of logistic elements and their interface with other functions of the firm. LSA data resulting from the LSA process is, in turn, stored as the logistic support analysis record (LSAR). As the logistics data base, the LSAR is the vehicle for communicating information for risk analysis, trade-off studies, effectiveness studies, and life cycle cost projections. A critical element of the communications model is a free and unrestricted two-way flow of information.
- D. The verification of supportability and logistic support. An LSA program,

through testing, verifies product supportability, the accuracy of LSA data, the adequacy of logistic products, and the achievement of logistic goals.

Successful achievement of each of these steps leads to achievement of the primary LSA objective and products that are affordable and capable of meeting operational requirements.

Process of LSA

The LSA program, to be effective, must be established and maintained as part of the overall integrated logistic support program. Only through the integration and co-ordination efforts of the ILS manager can the diverse disciplines making up the logistic elements merge into a united whole capable of meeting program objectives. LSA must be planned, developed, integrated, and implemented with product design, development, and production if it is to function effectively. It must include both management and technical resources for the successful performance of LSA tasks and the achievement of LSA requirements.

The LSA process is iterative in nature and can be divided into two general areas of activity: (1) support analysis and (2) supportability assessment and verification analysis. These two elements of the LSA process are related to the product design and development phases as illustrated in Figure 14.1.

Support Analysis

The *support analysis* phase of the LSA process is initiated at the system level to influence product design and operating concepts. In support of this activity, the LSA process identifies the gross logistic resource requirements of alternate approaches and analyzes the relationship between product design, operation, personnel requirements, and support characteristics to product objectives and goals. System-level support analysis is characterized by

A. The analysis of existing systems that are comparable to the product being developed
B. The determination and subsequent analysis of personnel, cost, and operational readiness drivers
C. The identification of potential targets of opportunity within the logistics support arena
D. The trade-off between alternative support concepts such as a built-in test capability as opposed to external support equipment

The completion of logistic goals at the system level is followed by a shift toward lower level ILS element analysis and refinement of the ILS system within the limits established during system-level analysis.

ILS element-level analysis emphasizes logistic support resource requirements

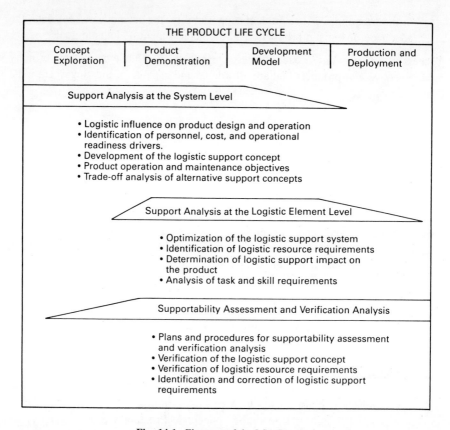

Fig. 14.1. Elements of the LSA Process

through an integrated evaluation of all product operation and maintenance functions. The objective is to determine task times, task frequencies, personnel and skill requirements, spares and repair part needs, test and support equipment requirements, and so forth—all functions, activities, and tasks that make up the logistic resource. The logistic support system is optimized through the allocation of functions and tasks to the various maintenance levels, repair-versus-discard analysis, formulation of recommended design changes, and related activities. The data developed during this phase of activity provides source information for each of the logistic elements.

Supportability Assessment and Verification Analysis

The supportability assessment and verification analysis phase of the LSA process extends throughout the product life cycle. The purpose of this phase of activity is to demonstrate, within established limits, the validity of support analysis and of the support products derived from that analysis. A secondary objective is to, as necessary, refine and adjust the analytical results and products.

Application of LSA

Logistic support analysis is directed toward the evaluation of a product in terms of its impact on logistic support resources. Analysis of this impact leads to logistic inputs into the design and development process and subsequent changes that enhance logistic supportability. In this manner, the LSA process is used to derive optimized solutions to a wide variety of problems endemic to logistic support of the new product. The application of LSA to these problems may be accomplished in several ways.

First, the LSA process assists the selection of alternatives. The product or system is designed to meet an objective which, in the vast majority of cases, may be attained through various alternative approaches. LSA data can be used to derive the logistic support costs associated with each alternative, thereby facilitating the selection of a preferred approach. LSA is also an invaluable aid in the development of an alternative maintenance policy when establishing repair practices that are consistent with the maintenance concept and offer increased product support. Alternative choices among various items of test and support equipment are also aided through the application of LSA. The measurement of product parameters, for instance, is normally accomplished by using a variety of items, each capable of performing the task. LSA provides the information that is needed to make an informed selection of one device over another.

LSA also provides a service in the evaluation of product characteristics such as reliability and maintainability. Product reliability, for example, may be increased through the use of HI-REL components (those components that have been subjected to stringent quality control standards during the manufacturing process, thereby greatly increasing their reliability), redundant components, or simplified designs. LSA provides the tools for evaluating each method, thereby having a direct impact on final product design.

LSA, as a tool for the evaluation of alternatives, leads to a total system that combines an improved product with an optimum package of logistic resources.

Logistic Support Analysis Record

A critical element contributing to the implementation of an effective LSA program is an organized and systematic method for recording analyses data. This function is fulfilled by the *logistic support analysis record* (LSAR). The LSAR serves as a central data base for the input, storage, update, and retrieval of LSA and LSA-related information. Benefits accruing from use of the LSAR include the following:

- A. Assurance of compatibility between logistic support requirements and ILS documents that identify and define maintenance instructions, skill requirements, training instructions, and allocation of maintenance tasks
- B. Elimination of separate analysis activities that yield the same type of data

C. Reduction in the number of data systems required to store and retrieve LSA data, which, in turn, avoids the problems encountered when attempting to maintain duplicate files

The LSAR is best suited to computer technology where summarizations of data permit ready access to logistic support requirements.

LSA and the Product

LSA and the inherent analytical process is best suited to the new product where logistic analysis may be conducted concurrent with and integrated into the design and development process.

The new product moves inexorably through several life cycle stages during transitions that begin with an idea and conclude with product retirement at the end of its useful life. The various life cycle stages are identified as

- A. Exploration of the concept
- B. Product design
- C. Demonstration and validation of the concept
- D. Product development
- E. Production
- F. Deployment
- G. Retirement

LSA is capable of providing immense benefits throughout much of the product life cycle when it is implemented during concept exploration and continued through deployment. Figure 14.2 depicts the data that are available from the LSA process for assistance in each of the logistic elements. The support analysis portion of LSA is most active during the first three stages, as shown in Figure 14.1, with a decreasing level of activity throughout the demonstration and development phases. Support assessment, on the other hand, begins later and continues throughout deployment. Deployment, which begins when the product is delivered to the customer, is the major portion of the product life cycle and is the phase that benefits most from an early expenditure of resources for LSA. The primary benefit is an increased life cycle at significantly lower support costs.

Measurement Techniques of LSA

LSA constitutes a composite of technologies that are used to define logistic support requirements and to transport these support needs into the product design and development process. The techniques include

- A. Logistic support models
- B. Life cycle cost analysis
- C. Maintenance engineering analysis

1. Levels of Maintenance

2. Maintenance Tasks by Level of Maintenance
- Task Time
- Task Sequence
- Task Frequency

3. Test and Support Equipment by Level of Maintenance
- Quantity
- Type
- Utilization Rate
- Cost

4. Spares and Repair Parts by Level of Maintenance
- Consumable Items
- Nonrepairable—
  Quantity and Type
- Repairable Items—
  Quantity and Type
- Replacement Interval
- Inventory Levels
- Safety Stock Levels
- High-Value Items
- The Provisioning Cycle
- Pipeline Time
- Shelf Life
- Availability
- Ordering Costs
- Inventory Costs
- Material Costs

5. Technical Publications
- Publication Requirements
- Data Collection
- Costs

6. Personnel and Training
- Staffing Levels
- Skill Requirements by Maintenance Levels
- Attrition Rates
- Learning Curves
- Training Requirements
- Training Courses
- Training Data
- Training Costs
- Training Equipment

7. Facilities
- Operating Requirements
- Maintenance Requirements
- Facility Layouts
- Storage Requirements
- Utility Requirements
- Capital Equipment
- Special Equipment
- Environmental Requirements (Clean Rooms, Shielding, etc.)
- Facility Costs

8. Transportation
- Equipment Requirements
- Quantities, Types, and Locations
- Packaging and Shipping Requirements
- Costs

9. Other Data
- Availability (A_i, A_a, and A_o)
- MTBF, MTTR
- Reliability
- System Effectiveness
- MDT
- MMH/OH
- Turn-Around-Times
- Dependability
- Maintenance Actions per Year
- Life Cycle Costs
- Failure Modes and Effects Analysis

Fig. 14.2. Summary of Logistic Support Analysis Data

Analysis begins with the recognition of a problem and a management commitment to provide the resources needed to achieve a solution. The commitment is followed by a clarification of the problem and the objectives that will be served by the solution. It is important to limit the scope of the problem to something that offers a reasonable potential for success. The large and complex problem should be reviewed with the objective of breaking it down into smaller problems that may be resolved more efficiently. A very important first step is to define the real problem and not a symptom of the problem. For example, the excessive failure rate of a product may be caused by a lack of trained personnel rather than a design defect.

Upon identification of the problem, the next step is to identify possible alternative solutions. The problem solver should look to external sources for these alternatives in all but the simplest problem. The use of techniques such as "brain storming" may be appropriate for the more serious or complex problems. The idea is to identify all potential solutions that may lead to a successful resolution. Once identified and listed, many of these potential solutions are rejected as not feasible for one reason or another. The remaining few represent legitimate alternatives and must be subjected to further analysis through the analytical tools available to the problem solver (logistic support models, life cycle costing, and maintenance engineering analysis).

Logistic Support Models

A *model* is a simplified representation of reality that addresses selected features of the problem being analyzed. It is a tool the problem solver (analyst) uses to evaluate the probable results of various alternatives. The model is not a direct source of answers; rather, it is a tool that provides the information needed to support the decision-making process.

The model represents the real world in a variety of ways. The specific method of representation is a function of the nature of the problem and may employ one or more analytical techniques in attempting to resolve the problem. The model may be quite simple or extremely complex, depending upon the quantity of variables that must be considered and the number of alternate approaches requiring evaluation. Regardless of model complexity, it should incorporate each of the following features:

A. The model should represent the dynamics of the system being evaluated. This representation should be easy to understand and manipulate, yet it should be close enough to the real world to provide usable results.
B. The model should have all salient features of the system and emphasize those system factors that are most relevant to the problem.
C. Models should be designed as simple as possible to facilitate their use. The elaborate or highly complex model may actually discourage its utilization as a problem-solving aid.
D. The design should allow easy modification or enhancement of the model as new information becomes available.

Models may be mathematical or strictly an analog representation of the system being evaluated. Mathematical models, while more abstract than other types, offer significant advantages to the problem solver. Chief among these is the ability to handle the problem as an entity and allow consideration of all major variables simultaneously. The mathematical model also permits a comparison of many possible solutions, provides a means of indicating cause and effect relationships, and facilitates the projection of future events. In the evaluation of problems, models employ techniques such as network analysis, Markov processes, Legrange equations, probability, statistical distributions, Monte Carlo sampling, linear programming, queueing theory, regression analysis, sequencing, and econometrics to name but a few. The problem solver has a lot of help as there are many models available that simulate a wide variety of characteristics.

Life Cycle Cost Analysis

Life cycle cost (LCC) analysis is a basic tool for use in the evaluation of logistic resource requirements. It is used in conjunction with other system parameters, such as effectiveness and technical performance, to determine cost effectiveness. LCC introduces the economic element necessary for a comparison of the various system design and support alternatives. It is unique in that LCC permits an assessment of the element of risk in the decision-making process. LCC enhances evaluation of the logistic system by

A. Defining areas of high support cost that result from design decisions
B. Permitting an evaluation of the cost associated with alternative support policies such as repair versus discard and spares level versus the probability of a stock-out
C. Defining the impact of operational requirements on alternative logistic support policies
D. Providing budget estimates for us in cost projections and financial planning
E. Providing data to support "make or buy" decisions

LCC may be applied to any phase of the product life cycle. It, unlike the other analytical tools of logistics, is not limited to the design and development phases of the product life cycle. Reliability or maintainability analysis, for example, can do little to increase the reliability or decrease the time to repair the existing product. The analysis can only determine what is there; it cannot change a design that has been implemented. LCC, on the other hand, can begin at any stage of the product life cycle and provide information relative to the cost of ownership from that time on. In a simplistic example, this is exactly what the owner of an older automobile is doing when trying to rationalize the expense of a new car. (Although it must be admitted that reason and logic may be somewhat clouded during this exercise.)

LCC is a versatile tool for the evaluation of logistic support resources. Its use, however, does require the development of specific input data. These data include

A. A definition of product objectives and operational requirements, including information such as product utilization, estimated operational life, operating locations, the product population, operating environment, and so forth
B. A definition of the product support concept such as the number of maintenance levels and locations, anticipated maintenance actions by level of maintenance, and so forth
C. The allocated or predicted reliability and maintainability factors
D. A definition or estimate of costs elements, by category, or significant factors contributing to the cost of each element, including elements such as operation and maintenance, personnel and training, spares and repair parts, inventory maintenance, test and support equipment, transportation and handling, technical publications, facilities, and so forth

LCC output products include the total cost of ownership, the costs per year, and a profile of the various cost elements. These data must be evaluated in relation to the total contribution and validity of input data. Major contributors or questionable areas should be subjected to additional evaluation (through sensitivity analysis, for example).

Maintenance Engineering Analysis

Maintenance engineering analysis (MEA) incorporates the activities of reliability, availability, and maintainability (RAM) analysis. This aspect of the LSA determines those product factors related to maintaining the product. Two of these factors, reliability and maintainability, are discussed at length in Chapter 4. MEA develops the data necessary to derive various system measurement parameters. These parameters are discussed in the following paragraphs.

A. Inherent availability (Ai). *Inherent availability* is the probability that a system, when used under stated conditions in an ideal support environment, will perform as intended at a given point in time.[1] Scheduled (preventive) maintenance actions, logistic supply time, and administrative downtime are excluded from calculations.

$$Ai = \frac{MTBF}{MTBF + \overline{Mct}} \qquad (14.1)$$

where *MTBF* represents the mean time between failure and equals

$$MTBF = \frac{1}{\lambda} \qquad (14.2)$$

[1] An ideal support environment is one which includes all necessary test and support equipment, spares and repair parts, properly trained personnel, and so forth.

$\overline{Mct}$ is the mean active unscheduled (corrective) maintenance time or the statistical mean for all corrective maintenance tasks. It is also referred to as the mean time to repair (MTTR). λ represents the expected number of failures for each hour of operation.

$$\overline{Mct} = \frac{\sum_{1}^{N} Mcti}{N} \qquad (14.3)$$

where: $Mcti$ represents individual corrective maintenance tasks and N equals the number of tasks.

B. Achieved availability (Aa). *Achieved availability* is the probability that a system, when used under stated conditions in an ideal environment, will perform as intended at a given point in time. This differs from Ai in that preventive maintenance time is included. Only logistic supply time and administrative downtime are excluded.

$$Aa = \frac{MTBM}{MTBM + \overline{M}} \qquad (14.4)$$

In Eq. (14.4), $MTBM$ represents mean time between maintenance, which includes scheduled (preventive) and unscheduled (corrective) maintenance, and equals:

$$MTBM = \frac{1}{\frac{1}{MTBMs} + \frac{1}{MTBMu}} \qquad (14.5)$$

where: $MTBMu$ is the mean time between unscheduled maintenance actions and $MTBMs$ is the mean time between scheduled maintenance actions.

In Eq. (14.4), $\overline{M}$ is the mean maintenance time and it is equal to

$$\overline{M} = \frac{\left(\overline{Mct}\right)\left(\frac{1}{MTBMu}\right) + \left(\overline{Mpt}\right)\left(\frac{1}{MTBMs}\right)}{\left(\frac{1}{MTBMu}\right) + \left(\frac{1}{MTBMs}\right)} \qquad (14.6)$$

where: $\overline{Mct}$ is the mean active unscheduled (corrective) maintenance time and $\overline{Mpt}$ is the mean active scheduled (preventive) maintenance time.

C. Operational availability (Ao). *Operational availability* is the probability that the system, when used under stated conditions in the actual operating environment, will perform as intended when called upon. It is equal to

$$Ao = \frac{MTBM + \text{Ready time}}{(MTBM + \text{Ready time}) + MDT} \qquad (14.7)$$

where: ready time is the time when the system is ready for use but not being utilized.

MDT (mean downtime) is the time the system is not in a condition to perform its intended function. This time includes active repair time, administrative downtime, and logistics supply time.

There are numerous other system measurement factors such as dependability (D), capability (C), system effectiveness (SE), and so forth. Availability measures were presented as samples of some of the more familiar units of measure. Additional excursions into the realm of MEA are, however, beyond the scope of this text.

LSA for the Deployed System

LSA is designed for implementation during the design and development phases of the new product. The support analysis phase of LSA is almost exclusively devoted to these phases of the product life cycle. LSA does not, however, end its involvement with the product upon delivery to the customer (deployment). The supportability assessment and verification analysis phase begins soon after support analysis to verify, refine, or change the results of this initial analysis. This phase should then continue throughout production and deployment. The objective of this continued involvement is ever-improving logistic support of the product.

The supportability and assessment phase of LSA can also be applied to the existing product toward this same objective. The currently fielded system may have been deployed without benefit of an LSA program. Does this lack of an LSA indicate that the systems have been deployed without due consideration being accorded logistic support? Most assuredly not. The system was deployed with some degree of logistic support provided by each of the applicable logistic elements, although the quality of this support is almost certainly less than it would have been with LSA. The various elements of logistics are essential, with or without an LSA program.

In the absence of LSA, logistic requirements are, of necessity, derived through trial and error or through the human judgment of skilled practitioners within each logistic element. Consider training wherein the experienced instructor reviews the product to be instructed and makes decisions based on human judgment regarding the probable tasks that require training. Obviously, an optimum training course is unlikely to result. Or consider spares and repair parts where the experienced provisioner reviews each item of the provisioning parts list (PPL) and determines what should be spared and the spares quantity through (1) judgment and past experience, (2) consultation with technical specialists, (3) extrapolation from similar systems, and (4) a review of applicable technical literature. Again, it is unlikely the system will be as efficient as it should be. In each instance, logistic support is provided to the product; however, the logistic support in these instances is based on data that are more of a "best guess," albeit a guess by experienced professionals operating within their individual areas of expertise. This "best guess" method of determining logistic support needs is unlikely to result in an optimized package of logistic resources.

Supportability assessment and verification analysis have the potential of dramatic improvements in product support. Benefits resulting from LSA involvement include

A. Identification of design defects that require unnecessary support requirements. These may have to be accepted, or they may be resolved through a relatively simple and cost-effective product modification.
B. Identification or verification of deficiencies in the previously established logistic support criteria.
C. Identification of the logistic support costs associated with product operation.
D. Establishment of a true baseline logistic support concept.
E. Life cycle support cost estimates.
F. Identification and definition of potential logistic problem areas.
G. Identification of the required maintenance skills and personnel staffing.
H. Identification of potential product enhancements to improve supportability.
I. Evaluation of the adequacy of the product support concept.
J. Evaluation of the support system capability.
K. Identification of the degree of achievement of logistic goals.

The LSA process, when appended with feedback data from deployed systems, provides the basis for continued improvement in all logistic elements. The most dramatic improvements in logistic supportability normally accrue from those products that were deployed without the benefit of an LSA program. It should be remembered, however, that this improvement is due to the lower level of support initially provided to these products.

SUMMARY

Logistic support analysis (LSA) is the analytical element of integrated logistic support. The LSA program, through a variety of analysis techniques and procedures, provides logistics with the tools necessary to influence product design and development in the direction of logistic supportability.

The LSA process is divided into two major areas of activity: (1) support analysis and (2) supportability assessment and verification analysis. Support analysis should be initiated concurrent with the concept-exploration phase of the product life cycle. This early involvement keeps logistics in step with design, thereby facilitating the implementation of logistic requirements into the evolving product. This phase continues through product design and the development of feasibility demonstration models. Support analysis activities begin to decrease as the product nears final design and enters into production.

Supportability assessment and verification analysis begin soon after the initiation of support analysis to refine and modify or change the preliminary results of early evaluation. This activity continues through product production and deployment with continuing evaluations of previously developed logistic support concepts. The objective of this activity is to develop a product characterized by an increasingly efficient and effective logistic support system at a decreasing cost of service.

LSA, as a logistic resource, is of immense value in the search for the best of several alternatives. It is used to derive the relative benefits associated with each potential choice, permitting a selection of the one that is preferred.

The results of LSA are maintained as an organized collection of data in the logistic support analysis record (LSAR). The LSAR represents a knowledge resource that incorporates the information needed to complete the logistic task. Use of this resource greatly reduces the problem of obtaining data that is necessary for the timely completion of the many logistic tasks. Cost reductions in this area partially offset the LSA cost.

The LSA uses a variety of techniques and procedures in the successful attainment of logistic objectives. These include logistic support models, life cycle costing, and maintenance engineering analysis. Models, by providing a representation of the real world, allow trade-off analyses on alternatives. Life cycle costing deals with the economic element by providing estimates of the total cost of ownership and cost profiles by year. Maintenance engineering analysis evaluates reliability and maintainability parameters and related factors such as product availability measures.

LSA, while of most value in support of the new product, may also be used in the evaluation of logistic support for existing products. Significant benefits may be realized, particularly for those systems that have been deployed without benefit of an LSA program.

QUESTIONS FOR REVIEW

1. What is the objective of logistic support analysis?
2. What is the LSA process? What are the two major areas of activity included therein?
3. Discuss the time phasing of the two major areas of LSA activity in relation to the product life cycle.
4. What benefit is derived from the use of LSA as a tool for the evaluation of alternatives?
5. Define the logistic support analysis record (LSAR). Which of the logistic elements are served by the LSAR?

6. Why is LSA best suited to the new product? How can it be of value to the existing product?
7. What benefit is derived from each of the three measurement techniques identified in the text?
8. What purpose is served by the logistic support model?
9. What is the difference between inherent availability (Ai) and achieved availability (Aa)?
10. Summarize the benefits of an LSA program for the deployed system.

15

LOGISTICS MANAGEMENT AND CONTROL

The integrated logistics support manager began to emerge as a recognized professional during the late 1970s. The growth and credibility of the ILS manager as a viable force within the firm accelerated with the increasing recognition of logistics and its potential impact on profitability.

ILS management is unique because the ILS manager must deal with a variety of professionals, many of which function outside his or her direct managerial control. For instance, the engineering manager manages engineers, the finance manager manages the accountants and financial planners, and the production manager manages the resources of production. The ILS manager, on the other hand, manages a diverse assortment of disciplines (the logistic elements), related only through the tenuous thread of integrated logistic support. His or her problems are compounded because many, if not all, of the logistic elements are the domain of other managers within the firm. This lack of explicit managerial control means that it can be very difficult to assemble a package of logistic resources that is capable of meeting product objectives. The intent of this chapter is to evaluate those difficulties and examine the management resources available to the ILS manager.

What Is Management?

A necessary first step in evaluating the problems facing the ILS manager is a review of management. What is this resource that is available to assist the ILS manager in the performance of his or her managerial responsibilities?

Managers give direction to the firm. They evaluate the goals of the firm, establish objectives that support those goals, and organize resources within the firm toward the successful attainment of those goals. Managers, regardless of the designation that precedes their title, face identical problems. They must organize work in the direction of optimized productivity and lead the human element toward achievement.

Management is the art of getting things done through others. Managers support the goals of the firm by arranging work for others, not in performing all the tasks themselves. In fact, the vast majority of employees in the firm are engaged in nonmanagerial work.

First-Line Management

The management task varies with the position of the manager in the firm. The first level of management (Figure 15.1) is the *first-line manager,* who is directly responsible for the production of the goods and services provided by the firm. This level represents the transition point between management and nonmanagement personnel. First-line managers work with nonmanagement subordinates toward the successful achievement of work-related objectives. They spend relatively little time with other managers or with customers of the firm.

For these reasons, the first-line manager requires a strong blending of both managerial and technical skills. He or she must have an advanced knowledge and understanding of the activities being performed by subordinates. Within the logistic function, first-line managers are those who are responsible for the products of each logistic element, e.g., the training manager, the facilities manager, the technical publications manager, and so forth. The larger organization may, however, elevate logistic element managers to the next higher level of management.

Middle Management

The *middle manager* is the buffer between top management and the first-line manager. The middle manager receives overall strategies and policies from top management and translates them into specific plans of action for implementation by the first-line manager. The middle manager spends the majority of time analyzing data, communicating with other managers, attending various meetings, preparing reports, and providing assistance to the first-line manager. In middle management, managerial skills are more important than technical skills.

Top Management

Top management is responsible for overall operations of the firm. They establish policies and function as representatives to the external environment of the firm (the community, other firms, and the government). Top management personnel spend most of their time with peers and individuals external to the firm.

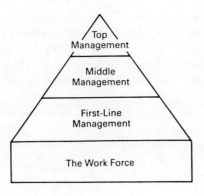

Fig. 15.1. Levels of Management

Management Thought

The approach to task accomplishment varies significantly with the skills, knowledge, attitudes, capabilities, and biases of the individual. Different methods and techniques are used by different individuals to perform the same task. Managers are no exception to this rule and are guided in the performance of their tasks by education, experience, and personal biases. For instance, they often adopt ideas and techniques from highly regarded peers and reject the management practices employed by those they regard less. All management techniques can, however, be traced back to one or more schools of management thought.

Traditional Viewpoint

The *traditional approach* to management focuses on a body of knowledge available to managers for creating stability and order within the firm. Traditional management places the emphasis on formal management processes such as planning, organizing, staffing, directing, and controlling. This approach tends to seek the one best way to manage and establish universal principles of management. Implicit within the traditional viewpoint is the simplistic assumption that the employee is motivated only by money or job security.

Traditional management, with its emphasis on formal procedures, the hierarchical structure, established authority, and vertical communications (the bureaucracy), generally does not represent the ideal environment for the ILS manager. The ILS manager has little formal authority (authority by virtue of position within the firm), since many logistics specialists "work for other managers." The resulting relationship requires a high degree of horizontal communications, which again violates the precepts of traditionalism.

Systems Viewpoint

The *systems viewpoint* of management looks at the firm as an accumulation of systems, each consisting of an input, a transformation process, and an output. (The concept of the firm as a system is treated at great length in Chapters 1 and 2.) Compared with other viewpoints, the systems viewpoint becomes more of a method for addressing management problems and issues than a school of management thought. The systems approach is useful because it prevents looking at problems as though there is a simple cause and effect relationship. Each problem is viewed as a system, with each element of the system exerting a greater or lesser impact on all other elements. This complex view of the managerial problem, unfortunately, does not provide concrete answers for dealing with these relationships. As a result, the systems approach to management requires a high degree of managerial skill. It, however, is a logical extension of the firm and the functions within the firm when viewed as a system. The ILS manager would be greatly benefitted by developing this managerial skill.

Behavioral Science Viewpoint

Behavioral science is the study of how people behave and why they behave as they do. This approach is important to the manager since only people have the ability to establish goals and initiate activity to achieve those goals. The application of behavioral science to management is, in general, applied from a systems point of view. It is more restrictive, however, because a strict behavioral science approach would limit its focus to the design of jobs used by the firm to reach defined objectives. This concept is important to the ILS manager in that an understanding of how people behave and why they behave as they do can be of great value in the absence of more formal authority.

Contingency Viewpoint

The *contingency approach* frankly admits that there is no one best way to manage. The "right" managerial technique varies with the individual, with the task, and with the situation. The contingency approach to management demands the development of conceptual skills. The manager must possess the ability to diagnose and understand the problem before tendering a solution.

Contingency management does not offer rigid rules of managerial conduct. Rather, the approach is to use a management style that is appropriate to the task, the situation, and the environment. For example, the new employee is faced with a strange environment and is normally uncertain about what is expected or even what the job is. The preferred managerial style in this case would lean toward close supervision and guidance in the performance of expected tasks, that is, the traditional approach. As this employee gains proficiency in job performance, close supervision and guidance are no longer required and would very likely be resented by the employee. The appropriate managerial style then becomes the more people-oriented approach—the behavioral science viewpoint.

Managerial Power and Influence

A central aspect of ILS managers' (and all managers') jobs is to influence others to act or behave in a certain way. The manager must influence the individual or the group toward the accomplishment of certain goals. The ability to influence is derived from power that is possessed by the manager; power and influence are the basic facts of life within the firm. The traditional approach awards power by virtue of position. The manager has the power to influence others based on a position within the hierarchical structure. This, however, represents only one source of power, one that is enjoyed in moderation, if at all, by the ILS manager.

Sources of Power

When the manager has the ability to influence others, it is inferred that he or she has power. Influence implies power, and power is necessary if the manager is to influence. Power, however, is much more than an ability to, through brute force, compel another to do one's bidding. Power exists in many forms, as the following list reveals:

A. Legitimate power. *Legitimate power* is based upon position in the hierarchical structure. This is the manager's formal authority. The follower responds to the position rather than the personal attributes of the individual in that position. Legitimate power has a long tradition in human history, and response of the follower to legitimte power is quasi-automatic. The king is obeyed because he is the king.

B. Reward power. *Reward power* is based upon the manager's ability to dispense rewards. Or, to state it more correctly, reward power is based on the follower's perception that the manager has the ability to dispense rewards. It must be recognized, however, that the reward must be something that is desired by the follower. The reward that is not desired by the follower will eliminate this as a source of power. The manager encountering this situation must use another source to influence behavior.

C. Coercive power. *Coercive power* is based on the follower's perception that the manager has the ability to inflict punishment. As with reward power, the punishment that is not feared will have little influence on behavior.

D. Referent power. *Referent power* is based upon the follower's desire to identify with the charismatic manager. This is power out of blind faith. Referent power is possessed by the great leader, the sports hero, the recording star, and other noted personalities whom followers desire to emulate.

E. Expert power. *Expert power* is based on the follower's perception that an individual has special knowledge or expertise that is needed by the follower.

The medical profession, for example, has expert power, and the physician has great power to influence behavior when medical knowledge has the potential of meeting the patient's need. Expert power is normally limited to a relatively narrow area of specialty. It is unlikely, for instance, that the medical doctor would possess expert power in the field of astronomy.
F. Representative power. *Representative power* is the democratic delegation of power to a leader for the purpose of serving the followers' interests. This source of power has very limited application in the firm.

Power by itself, however, is not adequate for the task. Coercive power, for example, will influence behavior only to the degree that punishment is avoided. This typically results in a relatively low level of performance, a level that is not acceptable in a competitive environment. Reward power also has disadvantages in that the employee soon grows to expect the reward, and increases in performance require increasing rewards. Another disadvantage, particularly for the ILS manager, is a lack of ability to dispense rewards. Reward power must be consistent if it is to influence behavior. The promised (or expected) reward that does not materialize eliminates this source of power. For these reasons, power must be supplemented by the ability to motivate. The employee must be motivated to perform.

Motivation and the ILS Manager

Motivation is the process that channels and sustains individual behavior. It is a subject of enormous importance to the manager because

A. The individual must be motivated to join the firm—and to remain with the firm
B. The employee must be motivated to (1) perform in the assigned role within the firm and (2) to exceed routine performance by engaging in creative and innovative behavior.

Motivation warrants a significant portion of managerial time and energy, since the motivated employee, more than any other factor, provides the firm with a competitive edge.

Process of Motivation

There are numerous theories concerning conditions in the work situation (other than skill and ability) that are associated with superior performance and employee satisfaction. Efforts in this area have led to two predominant theories. The first and most common is motivation theory, and the second is behaviorism. Each theory accounts for behavior in different ways. *Motivation theory,* for instance, is directed to the inner needs of people and makes assumptions involving their mental processes and needs. It defines behavior in terms of the individual's inner motivations. *Behaviorism,* on the other hand, is directed to the external needs of the individual. It ignores the concept of inner needs and defines behavior as a consequence of the work situation.

Motivation theory is subdivided into two distinct types of theories: Process theories ask how the individual is motivated, and content theories ask what motivates the individual. There is only one basic theory of behaviorism. A detailed treatment of these theories exceeds the scope of this text; however, the importance of this subject to the ILS manager warrants a brief discussion. Accordingly, this chapter briefly discusses two theories: Expectancy theory (one of the process theories of motivation) and behaviorism.

Expectancy Theory

Expectancy theory is the primary exponent of the process theory of motivation. It focuses on the ideas people have about their jobs and how these ideas combine with the desires of the individual to produce motivation. Specifically, expectancy relates to the belief of the individual regarding

A. An action that is expected to lead to a certain outcome or result
B. The valence or strength of desire for a particular outcome
C. The degree to which one outcome is expected to lead to the attainment of another outcome, called *instrumentality*
D. The estimate of the probability that applied effort will lead to an outcome, called *subjective probability*

The elements of expectancy theory cannot be considered in isolation. For example, the individual may believe with absolute certainty that performance will lead to a specific result. This will have no impact on behavior (performance) if the result is not desired (has a low valence). Or, to continue with this example, the fact that performance will lead to the result may be accepted, and the result is highly desired by the employee. Performance may still remain unchanged if the employee perceives a very low probability (low subjective probability) of attaining the desired goal.

The ILS manager and the manager who supervises one of the logistic elements must recognize these factors and their effects on motivation. The job task must be capable of fulfilling a need within the employee if it is to motivate. The manager must recognize this need, whether it be praise for good work, increased responsibility, peer recognition, growth within the firm, or another uniquely personal desire, and provide the work environment that is best suited to its achievement. This does not imply that the manager should attempt to satisfy a need for growth within the firm by promising a promotion to top management. The manager should, in this instance, discuss a career path with the employee that leads in that direction and offer encouragement regardless of the probability of success.

Behaviorism

Behaviorism is an alternative approach to understanding and controlling behavior. It does not consider the inner needs of the individual that motivation theory uses to explain why people act as they do. Behaviorists believe that behavior is the result of consequence. They also believe that behavior is learned through reinforcement. *Reinforcement* refers to the consequence of a behavior and can be either positive or negative. For example, spanking a child for lying is a negative reinforcement intended to teach that child not to tell lies. Spanking is the consequence that follows behavior, and a relationship is established between the two. Positive reinforcement consists of things that are good, pleasant, or satisfying as perceived by the individual. Negative reinforcement, on the other hand, consists of things that are perceived as bad, unpleasant, or unsatisfying. Repetitive reinforcement, positive or negative, leads to learned behavior.

Behavior may also be modified by the lack of reinforcement. This is called *omission*.[1] Omission destroys the relationship between behavior and the results of that behavior. This, over time, will cause the behavior to disappear since it no longer has any effect on the environment. The reverse of omission is punishment. This replaces a reinforcement with punishment, thereby leading to a behavioral change.

From the point of view of the manager, the differences between the two theories are not that significant since the management objective is to improve performance in the work place. Differences that do exist can become very difficult to understand, since both schools of thought frequently use the same event in attempting to prove the validity of their position. As an example, consider the employee whose performance improves following praise for good work. Motivation theories would present the position that praise fulfilled an inner need, which led to the improvement in performance. Behaviorists, on the other hand, would equate the praise to positive reinforcement and this is what created the change (improved performance). Who is correct? Does it really matter as long as the result is an improvement in performance?

Logistics Management Philosophy

The ILS manager is concerned with the allocation of resources and control of the logistic system. Successful accomplishment of these objectives requires the adoption of a management philosophy that complements the nature of logistic elements dispersed in the typical firm. One of these philosophies is called *management by objective*.

---

[1] Omission extinguishes a behavior pattern by withholding reinforcement, either positive or negative. For example, assume that praise (positive reinforcement) has been used to establish a behavior of "good" performance. Withholding (omission) of praise will destroy the relationship between good performance and praise and cause the learned behavior to disappear.

Management by Objective

Management by objective (MBO) is a philosophy of management that serves equally well as a way of life within the firm and as an aid to planning. As way of management life, MBO is a philosophy that can integrate all planning processes and day-to-day decision-making into a unified whole that benefits the firm. As a planning aid, MBO is a vehicle for translating management strategies into operating objectives and action plans through all levels and functions of the firm. The goals of MBO may be summarized as follows:

A. Achievement of a mutual problem-solving capability among levels of the firm when establishing objectives
B. Formation of open and trusting communications
C. Creation of win-win situations in all relationships
D. Rewards that are based on job-related performance
E. Minimization of office politics and the use of fear or force
F. Development of a supportive and challenging environment within the firm

This list clearly illustrates that MBO is a positive management philosophy that leans toward Theory Y as opposed to Theory X. Theory X and Theory Y are two contrasting management styles. They were first presented by Douglas McGregor in his book, *The Human Side of Enterprise,* published by McGraw-Hill in 1960. Theory X assumes the employee dislikes work and must be closely monitored and managed throughout the workday. Theory Y, on the other hand, assumes that the employee enjoys work, is self-motivated, and thus does not need an autocratic management style.

Process of MBO

The unique feature of MBO is the establishment and linking of objectives between top management and all functions within the firm. This process is extended to the individual employee whenever possible. It begins with top management, which establishes general, long-range objectives for the firm through consideration of the external environment of the firm and input from lower level managers. Objectives at this level are concerned with what the firm is and what it should be in the future.

Middle management quantifies the objectives and breaks them down into smaller objectives and operating plans appropriate to functions within the firm. These objectives and plans must support and complement the broader objectives of top management and the capabilities of the firm. The objectives and plans are then presented to first-line managers within each function of the firm where department and individual objectives are established.

The process of MBO is characterized by interaction between all levels of management. It cannot be implemented by top management and passed down to lower levels. Numerous attempts to implement MBO through such a top-down process have ended in failure. MBO, to be effective, must be implemented through a negotiated process where the subordinate and the supervisor, regardless of level, mutually agree on the objectives. The important factor to be considered is that each objective support and enable the overall objective of the firm.

Control of the Logistic System

Strategic plans are used to develop *operational plans*, which, in turn, provide the basis for action plans. *Action plans* define the activity that must take place to achieve the operational plan. Therefore, action plans require the allocation of resources within the logistic element in order to implement activities that are necessary for the successful achievement of objectives. Implicit within this series of events is the requirement for a method of controlling the logistic system. Logistic control is, of necessity, limited to management by exception.[2] The diverse nature of logistics requires this limiting of managerial attention to events that deviate from preestablished standards.

The ILS manager is ultimately responsible for the successful achievement of assigned objectives within each of the logistic elements. In meeting this responsibility, he or she must be able to manage and control elements within the logistic system. This ability can be derived through a system that enables the measurement of performance and an assessment of progress throughout the work cycle. A system for performance measurement allows early identification of potential problem areas, permitting timely management intervention and implementation of corrective measures. Thus, potentially serious problems can frequently be reduced or eliminated. Such a control system must, however, be flexible, since the degree of management control varies as a function of the cost of the activity being performed. There are significant differences in the management requirements between a $5 thousand repair contract and a $5 million request for a package of logistic resources. The effective manager cannot limit his or her activities to the management and control of resources within the firm. The management task must be expanded to incorporate a means to identify problems, both potential and actual, that require the application of management resources. Measurements provide this identification.

Typical measurement techniques rely upon monitoring the expenditure of resources over time. This concept is illustrated in Figure 15.2. This figure illustrates a partially completed program wherein current expenditures (the time-now line) are clearly less than projected expenditures. What else is revealed by this figure? Does the line AB represent a behind-schedule condition? Possibly, al-

---

[2]Management by exception establishes a tolerance zone or standards appropriate to each managed activity. Management intervention is then limited to activities or events that exceed those standards. For example: Management may elect to accept up to 24 hours of sick leave per employee per quarter (3 months). Absence in excess of this will initiate management action to evaluate the validity of excess sick leave.

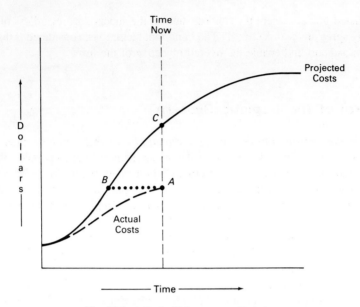

Fig. 15.2. A Typical Measurement Technique

though it may also represent a program that is ahead of schedule because more work was accomplished at less cost than anticipated.

The same type of reasoning also applies to line AC. It may represent a potential cost underrun. It is equally true, however, that it could represent decreased expenditures because of a lack of significant progress toward the logistic goals, leading perhaps to a cost overrun upon completion of the activity. In short, this measurement reveals absolutely nothing beyond the fact that actual expenditures are less than (equal to, or greater than) projected expenditures. A system that only measures expenditure over time gives no indication of accomplishment.

Earned Value

An acceptable management and control technique must measure accomplishment as well as the expenditure of financial resources. It must establish a plan of action and then measure actual work accomplished against both the plan and the expenditure of financial resources. The earned value concept meets these requirements.

To illustrate the concept of earned value, consider the logistic element spares and repair parts. Assume that this element has been tasked with the acquisition and delivery of twelve identical items over a period of one year. Further assume that a budget of $12,000 has been established for this activity. Given these assumptions, the first step in developing a management and control program is to establish a plan. The plan could state that twelve items would be delivered one year from the beginning of activity. This is certainly a plan, but is it a good one? To answer this question, consider a contract that is awarded for the construction of a house. Would the prudent individual transfer a large sum of money to the contractor, return one year later, and expect it to be completed and ready for occupancy? Certainly not! This is not a viable plan, since it does not provide for incremental measurement of progress (milestones) toward the ultimate objective.

The preferred plan designates incremental milestones wherein progress leading to project completion is measured. To illustrate this, assume a plan where the logistic element agrees, in conjunction with the customer, who requested the service, to acquire and deliver one item per month for a total of twelve months. Under this plan, twelve items will have been delivered at the end of one year, and the objective of the activity will have been fulfilled. The next step is to assign a dollar value to each milestone that is to be measured as progress toward the ultimate objective.

The total value of this activity is $12,000. Therefore, it appears reasonable to divide this amount equally among the twelve items to be delivered. This results in a value of $1,000 being assigned to each of the twelve milestones. The resulting expenditure rate is $1,000 per month. This is referred to as the *budgeted cost of work scheduled* (BCWS). The total value of all milestones is therefore equal to the value of the entire activity ($12,000). This process is illustrated in Figure 15.3.

Why is each milestone equal to $1,000? Because the plan stated that this was to be the value of each milestone. It should be apparent that individual milestones may be assigned any value as long as the total value of all milestones equals the value of the activity. Milestone value is actually a function of the cost of labor plus the cost of any purchased items that may be necessary for milestone completion. Since the original assumption assumed an equality of items, a linear distribution of the total value is reasonable.

The value of each milestone represents its *earned value*. This is the value that is earned by the performer upon completion of that milestone. Earned value is also referred to as the *budgeted cost of work performed* (BCWP).

Progress Measurement

Reporting on the logistic element at the end of each month (or accounting period) provides the ILS manager with a continuing measure of progress. This can be illustrated by the assumption of partial completion following an interval of time using the same example as above. Assume that (A) a total of five items have been acquired and delivered following five months of activity, and (B) actual costs total $5,500 during this five-month interval. (The *actual cost of work performed* (ACWP) is provided by the finance function of the firm.)

Several elements are revealed when these data are analyzed. First, the BCWP (earned value) is $5,000 since five milestones have been completed, each having a value of $1,000. Note that BCWP is independent of time. It is the aggregate value of each completed milestone, regardless of the time required for completion.

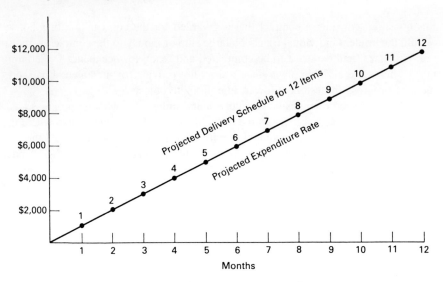

Fig. 15.3. A Plan for Task Accomplishment

Second, there is a negative *cost variance* (CV) of $500. The cost variance is calculated in Eq. (15.1).

$$CV = BCWP - ACWP \quad (15.1)$$

Substituting the example data in this equation yields

$$CV = \$5,000 - \$5,500$$
$$= -\$500$$

Does the negative cost variance indicate a cost overrun at completion? Not necessarily. A negative variance simply indicates that actual expenditures have exceeded projected expenditures. The cost variance does, however, provide the ILS manager with an indication that an out-of-plan condition exists. This should be investigated to determine if it does, in fact, represent a potential problem that should be corrected.

The counterpart of CV is the *schedule variance* (SV). SV is calculated in Eq. (15.2).

$$SV = BCWP - BCWS \quad (15.2)$$

BCWS is equal to $5,000, which is the projected expenditure rate of $1,000 per month times the five months of activity. Substituting this value, and the BCWP of $5,000 into Eq. (15.2) yields

$$SV = \$5,000 - \$5,000$$
$$= 0$$

The SV equal to zero indicates that all planned work has, in fact, been accomplished.

Continuing with this example, assume that eight months of activity have resulted in the acquisition and delivery of seven items at actual expenditures of $6,800. BCWS now equals $8,000; BCWP, $7,000; and ACWP, $6,800.

The resulting variances are

$$SV = BCWP - BCWS$$
$$= \$7,000 - \$8,000$$
$$= -\$1,000$$

and

$$CV = BCWP - ACWP$$
$$= \$7,000 - \$6,800$$
$$= \$200$$

Once again, the variances indicate a deviation from the plan. They should only be regarded as signals to initiate an investigation. The potential seriousness of a variance does, however, increase as the activity nears its scheduled completion date. The magnitude of the variance and its value relative to the total budget are also indicators of potential seriousness. The variances and the three performance measures (BCWS, BCWP, and ACWP) may be calculated for each period or on a cumulative basis.

Purchased Item Measurement

The logistics function, particularly the spares and repair parts element, expends a significant amount of resources in the purchase of items from sources external to the firm. This presents a unique problem in earned value measurement, because performance is, in part, determined by events that are beyond the control of the logistics practitioner. The logistic element is responsible for identifying the need to purchase an item and for ordering the item (or for seeing that it is ordered) from a supplier of the firm. Receipt of the item by the firm is, however, under the control of the supplier and associated transportation resources. Monies allocated to purchase items should be distributed throughout the performance period as a function of planned payments to the supplier. Planned expenditures, based on the plan for payment by the finance function, constitute the BCWS. BCWP is then based on estimates of when the item is *expected* to be in the firm and available for use. ACWP is, of course, actual payment by the finance function.

BCWP (earned value) is purposely claimed when the item is available for use, not necessarily when it is actually received by the firm. The reasoning behind this is that the time interval between its arrival at the receiving dock and its availability for use approximates the time required between receipt and payment. This keeps BCWS, BCWP, and ACWP in synchronization.

The use of earned value with purchased items is one of the problems associated with this concept. The example above represents only one way of attempting to measure performance in this area. Many others have been developed, with each attempting to attain a more realistic match between the three performance measures. All are rough approximations at best, and the preferred method is largely a matter of management choice.

Milestone Selection

An effective performance measurement system demands the inclusion of incremental milestones as positive indicators of progress. Milestones should be selected by the logistic practitioner who is responsible for performing the activity, subject to approval by the logistic element supervisor and the ILS manager. The approval cycle assures that the milestone represents a realistic and effective measure of progress. A good milestone possesses the following characteristics:

A. It is a product or an event.
B. It is clearly within the authority of the performer.
C. It possesses clear and objective criteria for measurement.
D. It has an assigned dollar value that is a portion of the total activity budget.

These criteria virtually eliminate the use of percentages as indicators of progress toward completion. A third party cannot define or quantify percent complete as a measure of progress. The milestone must be something that can be observed, weighed, counted, or otherwise verified through an objective standard.

Discrete milestones, while providing the most meaningful measure, may not be possible for each job task element. Accordingly, the methods listed below for determining earned value may be used in specific instances.

A. The 50-50 technique. This method is reserved for the task that begins in one reporting period and is completed in the succeeding period. One-half, or 50 percent, of the value is claimed as BCWP in the initial period. The remaining BCWP is claimed in the next reporting period (assuming task completion).
Other ratios, such as 40-60 or 30-70, may also be used with this technique.
B. The 0-100 Technique. This method is a special case of the 50-50 technique. It is used for the task that begins and ends within a single reporting period. No BCWP is claimed when the activity is initiated; instead, the total milestone value is claimed as BCWP upon completion.
C. Level of Effort. Selected activities such as managing the logistic task do not lend themselves to the establishment of milestones. While it is not a preferred method of measure, these activities may be measured by level of effort. When using level of effort, BCWP is, by definition, equal to BCWS. This means that the schedule variance is always zero. Level of effort can, however, exhibit a cost variance. For example, the BCWS for level of ef-

fort measurement is based upon the supervisor's estimate of the time that will be required in managing the logistic task. ACWP, as recorded by the finance function, is the dollar value of the time actually spent in this management activity. Any difference between the two constitutes a cost variance.

It is important to realize that level of effort is a means of measuring performance. It does not imply that an equal distribution of time (or the dollar equivalent of that time) is scheduled for each reporting period.

There are other methods, such as apportioned effort, item count, and so forth, that may be used to evaluate progress toward an assigned objective. The preceding list does, however, identify the most common.

Evaluation of the Report

Data resulting from the performance measurement system provide the logistic element manager and the ILS manager with a source of information for managerial decision making. The data reflect information on the current period and cumulative information. Care must be exercised when evaluating current period information, since it is much more volatile than aggregate (cumulative) information. This is a result of the assigned milestone value and the necessity of reporting at a given point in time.

Consider the two activity periods illustrated in Figure 15.4. In this illustration, milestone 1 has slipped from the latter part of period 1 into the first part of period 2. This slip is indicated by the dashed line extending from the originally scheduled milestone (the triangle) in period 1 to the rescheduled milestone depicted in period 2 (the diamond). Note that this slip could be for as little as one day if the completion date was projected as the last day of the reporting period.

The lack of milestone completion during period 1 means that no earned value (BCWP) may be claimed. BCWS, however, is $5,000 and, as work was performed toward completion of this milestone, costs were accumulated against the activity. ACWP, as revealed by the finance function of the firm, is $4,800. This results in a relatively high negative variance in both cost and schedule.

Period 2 assumes completion of both the rescheduled milestone from period 1 and milestone 2, which is scheduled for completion during this period. Earned value is claimed for both completed milestones leading to a period 2 BCWP of $12,000. Period 2 BCWS and ACWP are both $7,000, which yields a large positive variance in both cost and schedule.

Cumulative variance, on the other hand, looks at the summation of both period 1 and period 2. This summation results in BCWS, BCWP, and ACWP all being equal to $12,000. This provides a cumulative cost variance and schedule variance that is equal to zero. Thus, cumulative variances are more meaningful than period variances in that they more accurately reflect the real situation.

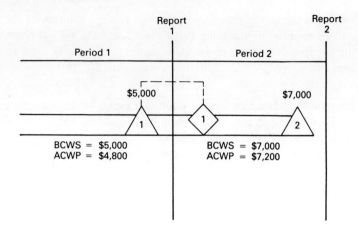

Fig. 15.4. Period versus Cumulative Data Analysis

Performance Indices

Performance indices permit the logistic manager to quickly evaluate both efficiency and performance. Efficiency measures include a cost and a schedule element. The *cost performance index (efficiency)* or CPIe and the *schedule performance index (efficiency)* or SPIe are calculated as follows:

$$CPIe = \frac{BCWP}{ACWP} \qquad (15.3)$$

$$SPIe = \frac{BCWP}{BCWS} \qquad (15.4)$$

The *cost performance index (performance)* or CPIp indicates the actual cost of each dollar of work accomplished.

$$CPIp = \frac{ACWP}{BCWP} \qquad (15.5)$$

These indices may be illustrated by using the performance measures developed for the concept of earned value. For month 8 of the sample problem, these measures were BCWS = $8,000; BCWP = $7,000; and ACWP = $6,800. Substituting these values in the performance indices equations yields

$$CPIe = \frac{\$7,000}{\$6,800}$$

$$\doteq 1.03$$

$$SPIe = \frac{\$7,000}{\$8,000}$$

$$\doteq 0.875$$

$$CPIp = \frac{\$6,800}{\$7,000}$$

$$\doteq 0.97$$

The efficiency measures tell the manager that past activities have been performed at 103 percent efficiency when measured against the ACWP and at 87.5 percent efficiency when measured against the BCWS. CPIp indicates that each dollar of earned value was obtained at an actual cost of 97 cents.

These measures relate to the performance of past activity. The counterpart for measuring activity in the future is the *to-complete-performance index* (TCPI) and it is measured as follows:

$$TCPI = \frac{\text{Work Remaining}}{\text{Money Remaining}} = \frac{BAC - BCWP_c}{EAC - ACWP_c} \quad (15.6)$$

where: *BAC* equals the activity budget
BCWPc equals the BCWP (cumulative)
ACWPc equals the ACWP (cumulative)
EAC equals the performer's estimate of the budget that will be required in completing the total activity. This may be equal to the original activity budget (on-budget), greater than (an overrun), or less than (an underrun).

Assuming that the BAC and EAC are both equal to the earned value example budget ($12,000) and using the same performance measures as before, the TCPI is calculated as follows:

$$TCPI = \frac{\$12,000 - \$7,000}{\$12,000 - \$6,800} = \frac{\$5,000}{\$5,200}$$

$$\doteq 0.96$$

This indicates that at least 96 percent efficiency is required in future performance if the activity is to be completed within budget.

Schedule Conversion

Schedule conversion techniques provide a means of converting report data to the number of months an activity is behind or ahead of schedule (X).

$$X = \frac{SVc}{\frac{BCWPc}{T}} \quad (15.7)$$

where: T is months of activity to date
SVc is schedule variance (cumulative)
BCWPc is budgeted cost of work performed (cumulative)

$$X = \frac{(BAC \times SPIp - BAC)}{BCWPp} \quad (15.8)$$

where: BAC Budget at Completion
SPI_p Schedule Performance Index (current period)
$BCWP_p$ Budgeted cost of work performed (current period)

$$X = \frac{SVc}{BCWSp} \quad (15.9)$$

where: $BCWS_p$ is Budgeted cost of work scheduled (current period).

Using Eq. (15.7) and the earned value example data as calculated using Eq's. 15.1 and 15.2 schedule conversion yields:

$$X = \frac{-1,000}{\frac{7,000}{8}} \neq \frac{-1,000}{875} = -1.1$$

$$\neq -1.1$$

This technique reveals that the activity is 1.1 months behind schedule.

These evaluation techniques provide, at best, an approximation of true program status. The manager should use them as tools to indicate potential problems. None of the tools, in and of themselves, provide a definitive analysis or point unerringly to a problem.

SUMMARY

The ILS manager is faced with a challenging task in that logistics is made up of a number of diverse disciplines as most of the logistic elements are the domain of other managers within the firm. This may lead to a lack of managerial control that makes it very difficult to assemble an optimum package of logistic resources.

places him or her in middle management, subordinate to top management and supervising the first-line manager. The first-line manager is normally the supervisor of the logistic element.

The manager approaches the task of management through a variety of methods and techniques. The traditional approach focuses on the formal aspects of management which are characterized by the hierarchical structure, established authority, and vertical communications. The systems approach to management on the other hand, views the firm as an accumulation of systems. This is more of an approach to problem solving than a school of management thought. This approach does, however, present an advantage to the ILS manager in that it reduces the tendency to look for simple cause and effect relationships.

The behavioral science approach to management focuses on the individual and the underlying reasons behind behavior. This viewpoint is valuable to the manager as only people make decisions and set goals.

The contingency approach differs in that it admits that there is no one best way to manage. The preferred style varies as a function of the individual, the situation, and the task.

The managerial task is to influence others through the intelligent use of power. Several sources of power are available to the manager such as legitimate, reward, coercive, referent, and expert. Power, however, is insufficient by itself. It must be used in conjunction with motivation that channels and sustains individual behavior.

Motivation follows two basic schools of thought: motivation theories, which are inner-directed, and behaviorism, which is outer-directed. An understanding of motivation is important to the manager, since employees can be motivated to be more productive, thereby providing the firm with a competitive edge.

Management must incorporate the element of control if it is to fulfill its objectives. Initial efforts at control were limited to the monitoring of expenditures over time. These efforts proved to be inadequate because they gave no indication of accomplishment. The concept of earned value, which measures cost in terms of accomplishment, can be added to the managerial control process. The earned value concept divides each activity into discrete events or milestones, with each being assigned a proportionate share of the total activity budget. Earned value claimed upon completion of the milestone measures performance through a comparison with the established plan and the actual incurred cost. The performance measures are BCWS, BCWP, and ACWP.

QUESTIONS FOR REVIEW

1. How does the diversity of logistic elements increase the difficulty of the ILS manager's task?
2. What is management? Differentiate between the levels of management.

3. What source of power is most appropriate for the ILS manager? What source is least appropriate?
4. Describe the expectancy theory of motivation.
5. What is management by objective?
6. How do the motivation theories differ from behaviorism?
7. What is wrong with a control process that monitors expenditures over time?
8. Descibe the concept of earned value.
9. Why are cumulative variances of more value to the manager than period variances?
10. What is the meaning of a TCPI of 1.07?

16

THE FUTURE OF LOGISTICS

What does the future hold for logistics and the ILS manager? Will logistics become a more cohesive and integrated structure within the firm? Will the ILS manager play a more important role in the corporate decision-making process? What skills should the logistics practitioner develop to enhance individual career potential? What educational opportunities are available to assist in this development? What tools will the logistics practitioner of tomorrow have to enhance job performance?

ILS managers have been searching for answers to these elusive questions for years with almost no results, perhaps because these questions are being asked in the wrong environment. The typical ILS manager thinks in terms of the logistic system, and discussions related to these questions are limited to an audience of peers. Agreement as to the nature of problems facing the logistics function and the solutions to those problems is naturally high, and the collective assembly of ILS managers enthusiastically agrees with the precepts set forth. But is this the correct forum for needed growth within the logistics arena? Other than an enormous boost to the ego, little of value transpires during a debate in which all involved hold the same interests, beliefs, and prejudices.

Integrated logistic support systems require the expenditure of large amounts of money and time with no return on investment until some time in the distant future. Consider the customer who purchases a product from the firm. The added cost in today's dollars of adding, for example, the supportability assessment and verification analysis portion of the LSA may represent a substantial increase in the product acquisition cost. The return on this investment, however, is realized over a period of years through improvements and refinements in logistic supportability. Returns over the product life cycle may eventually exceed $100 for each dollar that is spent on providing this element of logistic support.

The ILS manager, while he or she may fully believe an investment in logistics is worthwhile, cannot authorize the commitment of corporate resources toward this end. This lack of capability dramatically indicates that the ILS manager is not the proper individual to address. A united front is critical and necessary. However, the role and potential of logistics must be communicated to the environment that exists beyond the logistics arena, that is, to those setting the overall objectives of the firm. The logistics practitioner is the one who must bear the responsibility for initiating and maintaining this communication.

Another difficulty facing the amalgamation of logistics into an integrated function within the firm is the typically dispersed nature of logistic elements. Total integration will remain a myth until the elements of logistics have been merged into a logistics function within the firm.

Inducements to Change

Logistics, as vital as it is to those who function within the field, must be viewed in the proper perspective. As important as it is, logistics is simply one of many subsets of the ultimate system—society. Logistics is a system dedicated to service and the creation and maintenance of a defined level of product support. The demand for this support continues to grow with changes in society. Logistics, in remaining a viable force, must change in concert with these changes in society.

Economic and political climates have undergone major changes in the recent past and are likely to continue to do so. These changes have significantly impacted the environment of the firm. Important factors contributing to these changes include

A. Demographics. The population is exploding in Third World nations and poorer parts of the world, while there is minimal growth in more developed nations. The resulting shift in population distribution will lead to a realignment of the product market as demand shifts toward less-developed countries. At the same time, the increased population will create vast new labor pools in these same countries. The resulting increase in the trend toward overseas production will require larger logistic systems and improved managerial control through better comunications.

B. Sociopolitical Changes. Society is becoming more "world-oriented," and worldwide political loyalties are decreasing in both scope and magnitude. Influences that have existed since World War II are being diluted as emerging political forces rise in countries such as Japan and Brazil. The gradual shift to a world economy will mean ever more complex logistic systems. Additional impacts are being experienced in the more-developed nations as special interest and consumer advocate groups increase their sphere of influence.

This has the potential of creating more government regulations and greater complexity for the logistics function.
C. Economics. There is a growing scarcity of raw materials and fuels produced by the extraction industries. This is the result of increasing demand and temporary shortages that are often caused by an unstable political climate. The oil crisis of the 1970s is a good example of the latter cause. This trend has created financial instability throughout the world, and it promises to continue.
D. Technology. An advantageous change is the accelerating use of the computer as a management tool. This, if nothing else, is allowing the manager to cope with a changing and more complex environment. Accelerating technology creates management problems of its own, however, as the problems of how and when to use the new item must be addressed.

The Solution

The systems approach is the cornerstone of an integrated logistic support concept. None can question the logic that integration presents significant advantages in the use of resources to accomplish objectives of the firm. Application, not the concept, is the problem. The typical firm pays lip service to the implementation of an integrated logistic system while establishing boundaries that further isolate the elements of logistics. This creates a situation where the analysis of trade-off potential is virtually impossible.

There are a variety of approaches described throughout this text that are available and fully capable of meeting the logistic challenge of tomorrow. These approaches exist within the philosophy of an integrated logistic support system and may be implemented using today's technology. The responsibility for this implementation resides jointly with the ILS manager and management of the firm. The ILS manager can communicate awareness, but only top management is capable of making the commitment of resources necessary for implementation.

Logistics must extend beyond the domain of the individual firm to embrace many other organizations. There must be an increased awareness of the systems approach wherein suppliers to the firm and customers of the firm are considered as an integrated whole.

A growing service-oriented economy foretells a demand for logistic services that will increase at a tremendous rate. The approach to meeting these awesome challenges resides in the philosophy of the system and integrated logistics support. And the time for its implementation is now!

Index

Index

A

Accuracy of the publication, 162
Achieved availability, 211
Acquisition of knowledge, 144
Acquisition cost, 57
Act to regulate commerce, 70
Actual cost of work performed, 227
ACWP, 227
Air transport, 67
Airtruck, 69
Appendix, 172
Arbitrary, 71
Asset deployment, 92
Audio-visual aids, 158
Availability, 133
  achieved, 211
  inherent, 210
  operational, 211
Average inventory, 97, 109
Average stock, 97
Average workload, 119

B

Back-haul, 29
Balance supply and demand, 94
Base stock, 97
Bathtub curve, 51
BCWP, 227
BCWS, 227
Before tax profit, 139
Behaviorism, 222
Bill of materials, 115
BIT, 196
BITE, 196
Budgeted cost of work performed, 227
Budgeted cost of work scheduled, 227
Built in test, 196
Built in test equipment, 196

243

244 / Index

C

Capability, 133
Capacity control, 106
Capacity planning, 106
Capacity requirements, 115, 119
Car load, 71
Carriers:
 common, 68
 contract, 68
 exempt, 68
 legal definition, 67
 private, 67
Channel arrangement, 6
Checklist, 151
CL, 71
Clarity of presentation, 162
Class rates, 70
Coercive power, 220
COFC, 69
Combination rates, 72
Commercial provisioning, 178
Commodity rates, 70, 71
Common carrier, 68
Common test and support equipment, 193
Competence, level of, 161
Concept exploration, 47
Conference, provisioning, 179
Conflict, 32
Consolidated shipments, 141
Container on flatcar, 69
Contingency management, 219
Contract carrier, 68
Corrective maintenance, 77, 195
Cost:
 fixed, 104
 variable, 104
Cost elements, inventory, 103
Cost performance index, 232
Cost variance, 228
Course length, 152
Course materials:
 audio-visual aids, 158
 course outline, 157
 examination, 158
 hands-on training guide, 157
 lesson guide, 157
 student text, 158
 training plan, 156
Course prerequisites, 155
CRP, 115, 119
Cumulative variance, 231
Customer:
 external, 130
 internal, 132
Customer of the firm, 132
Customer service, 6
Customer service policy, 135
CV, 228

D

Damage induced failure, 196
Data base management system, 154
DBMS, 154
Decoupling, 94
Defense Logistics Agency, 177
Demand, 109
Demand pull, 122
Department of Defense, 177
Department of Defense Instruction 3232.7, 177
Dependent failure, 195
Depot maintenance, 56, 194
Derating, 50
Discrete lot sizing, 110
Distribution channel, 6
Distribution requirements planning, 114
DLA, 177
DODI 3232.7, 177
DODI 4140.42, 178
DRP, 114
DRP II, 114

E

Earned value, 226
Economic justification, 86
Economic order quantity, 107
Education, 144
EERC code, 184
Effectiveness, 37
Efficiency, 37
Elements of logistics, 17
Elkins Act, 69
Emergency Transportation Act, 70
Enabling objectives, 149
Entrance criteria, 155
Entry level, 155
Environment:
  general, 33
  task, 33
EOQ, 107
Examinations, 158
Exception rates, 70
Exempt carrier, 68
Expectancy theory, 222
Expected repair actions, 77
Expected workload, 76
Expert power, 220
External customer, 130

F

Facilities, 11
Facility size, 81
Facility workload, 78
Failure rate, 51, 183
Failure rate curve, 51
Failures:
  damage induced, 196
  dependent, 195
  maintenance induced, 195
  operator induced, 195
  primary, 195
  wear-out, 196

FAK, 72
Federal Aid Highway Act, 64
Finance, 18
First line management, 217
Fishyback, 69
Fixed cost, 104
f.o.b. destination, 97
f.o.b. origin, 97
Form postponement, 141
Forms of inventory, 93
Forward, 171
Freight of all kinds, 72
Freight rates, 71
FriendlySoft, 132
Front matter, 171
Functions of inventory, 94
Functions of the firm, 30

G

General environment, 33
Geographic specialization, 94
Government provisioning, 178
Gross maintenance time, 54
Growth phase, 38
Guidance conference, 170

H

Handling activities, 84
Hands-on training guide, 157
Hepburn Act, 70
Hi Rel, 50
Hub and spoke, 67
Human engineering, 149

I

ICC, 69
ILS manager, 4, 47
Incremental cost, 57

Index, 172
  cost, 232
  schedule, 232
  to complete performance, 233
Infant mortality, 51
Inherent availability, 210
Instructor, logistics, 145
Instructor preparation, 155
Instrumentality, 222
Intended audience, 163
Interdependency triangle, 61
Intermediary specialist, 22, 34, 83
Intermediate maintenance, 56, 194
Intermediate positioned warehouse, 86
Internal customer, 132
Internal inventory transfer, 9, 129
Interstate Commerce Commission, 69
Interstate highway system, 64
Interview, 151
Introduction, 171
Inventory, 29, 92
  assortment, 91
  average, 109
  cost elements, 103
  cycle, 96
  cycle frequency, 100
  forms of, 93
  function of, 94
  management, 91
  marketing, 93
  materials, 92
  obsolescence cost, 104
  opportunity cost, 104
  order cost, 104
  performance cycle, 98
  pipeline, 97, 101
  product, 93
  spare and repair parts, 92
  storage costs, 103
  transfer cycle, 36
  transportation cost, 101

J

JIT, 22, 121
Job analysis, 165
  priority sequence, 166
Joint rates, 72
Just-in-time, 22, 121

K

KANBAN, 122, 125
  move card, 125
  production card, 125

L

LCC, 4, 48, 57, 209
LCL, 71
Lead time, 109
Least total cost design, 136
Legitimate power, 220
Less-than-carload, 71
Less-than-truckload 71
Lesson guide, 157
Lesson objective, 149
Level of competence, 161
Level of effort, 230
Level of repair analysis, 56
Levels of maintenance, 55
Life cycle cost, 4, 48, 57, 209
Literacy, 161
Local rates, 72
Location planning, 85
Logistic:
  availability, 133
  capability, 133
  channel, 34
  coordination, 8

elements, 17
facilities, 25, 74
operations, 8, 35, 40
performance, 128, 133
performance cycle, 36, 133
service, 128
specialist, 34
support analysis, 178, 181, 201
support analysis record, 181, 201, 205
support models, 208
Logistics:
definition, 1–2
DOD definition, 3
function, 20
horizontal array, 15
instruction, 146
instructor, 145
management, 3
mission of, 11
pyramid of subsets, 16
of risk, 91
support function, 16
training element, 144
Logistics Direction, 121–150, 177
LORA, 56
Lot sizing, 110
Lot-for-lot, 111
LSA, 178, 181, 201
applications, 205
deployed system, 212
objectives, 202
process, 204
support analysis, 203
supportability assessment, 204
verification analysis, 204
LSAR, 181, 201, 205
LTL, 71

M

Maintainability, 48, 52–53
Maintainability engineering, 79

Maintenance, 52
activity, 76
corrective, 77, 195
depot, 56, 194
facility, 75
intermediate, 56, 194
levels of, 55
organizational, 55, 193
predictions, 78
preventive, 77, 123, 194
scheduled, 195
training, 147
unscheduled, 195
Maintenance engineering analysis, 210
Maintenance factor, 78
Maintenance induced failures, 195
Maintenance man hours, 53
Maintenance plan, 196
Management, 217
behavioral science view, 219
contingency view, 219
first-line 217
middle, 217
by objective, 223
systems view, 219
thought, 218
top, 217
traditional view, 218
Manipulative skills, 147
Manufacturing:
cycle times, 124
defects, 195
management, 123
risk, 95
work force, 129
Manufacturing resource planning, 10
Market based economy, 37
Market forecasting, 41
Market inventory, 93
Market positioned warehouse, 86
Marketing, 18
Marketing concept, 37
Master production schedule, 115, 118
Material:

Material (*cont.*)
  cycle, 36
  handling, 88
  management, 42, 46, 129
  requirements planning, 10, 111
  storage, 36
Materials inventory, 92
MBO, 224
MEA, 210
Mean downtime, 53
Mean time between corrective maintenance, 78
Mean time between demand, 187
Mean time between failure, 49, 76, 183
Mean time between maintenance actions, 78
Mean time between unscheduled maintenance actions, 18
Mean time to repair, 53
Middle management, 217
MIL-STD-1552, 178
MIL-STD-1561, 178
Milestones, 230
Mission of logistics, 11
MMH/OH, 53
Motivation theory, 221
Motor Carrier Act of 1980, 68, 70
Motor carriers, 64
Move card, 125
MPS, 111, 114–15, 118
MRP, 10, 111
MRP II, 114
MTBCM, 78
MTBD, 187
MTBF, 49, 76, 183
MTBMA, 78
MTBUMA, 187
MTTR, 53
Multimode systems, 69

N

Net maintenance time, 54

O

Objectives:
  course, 149
  enabling, 149
  lesson, 149
  of LSA, 202
  task oriented, 148
  terminal, 149
Observation, 152
Obsolescence cost, 104
Occupational survey, 151
Omission, 223
Operating hours, 53
Operation training, 147
Operational availability, 211
Operational plans, 225
Operational scenario, 76
Operations logistics, 2
Opportunity cost, 104
Optimum repair level analysis, 56
Order cost, 104
Order point, 99, 109
Order quantity, 109
Organizational maintenance, 55, 193
ORLA, 56
Out-of-production, 188
Overstocking, 91

P

Parallel relationship, 51
Peculiar test and support equipment, 193
Performance as expected, 49
Performance cycle:
  logistic, 128, 133
  training, 145
Performance cycles, 36
Performance indices, 232
Performance quality, 134
Period order quantity, 111
Period variance, 231

Personnel, 11
Personnel requirements, 80–81
Physical distribution, 129
Piggyback, 69
Pipeline, 67
Pipeline inventory, 97, 101
Place utility, 83
POQ, 111
Post production:
  planning, 189
  spares, 189
  support, 188
Postponement:
  form, 141
  temporal, 141
Power:
  coercive, 220
  expert, 220
  legitimate, 220
  referent, 220
  representative, 221
  reward, 220
PPL, 179
Preface, 171
Prerequisites to training, 147
Preventive maintenance, 77, 123, 194
Primary failures, 195
Priority control, 106
Priority planning, 106
Private carriers, 68
Probability, 48
Process of LSA, 203
Product:
  cycle, 36
  decline, 38
  distribution, 35
  introduction, 38
  inventory, 93
  life cycle, 38
  maintainability, 149
  market forecasting, 9
  phase-out, 38
  support, 131

Production, 18
Production card, 125
Production equipment, 129
Production and inventory control 105
Production positioned warehouse, 85
Proportional rates, 72
Provisioning, 175–78
  commercial, 178
  conference, 179
  documentation, 178
  government, 178
  history, 177
  process, 179
Provisioning parts list, 179
Public warehouse, 83
Publication development, 168
Publication front matter, 171
Pull system, 105, 122
Push system, 105, 122

Q

Quality, 124, 134
Questionnaire, 151

R

Rail miles, 66
Railroads, 66
RAM, 150
RAM analysis, 210
Rating, 71
Readability, 162
Redundancy, 50
References, 172
Referent power, 220
Reinforcement, 223
Reliability, 47–8
Reliability, tools of, 50
Repair level analysis, 48, 54
Repair level decisions, 56
Repair personnel, 80

Repair vs. discard, 55
Repairable items, 79
Representative power, 221
Request for proposal, 8
Retail outlets, 83
Retail risk, 96
Return on sales, 139
Reverse logistics, 22, 142
Reward power, 220
RFP, 8
Risk:
   manufacturing 95
   retail, 96
   wholesale, 96
Road transportation, 64

S

Safety stock, 43, 95–97, 109
Schedule conversion, 234
Schedule performance index, 232
Schedule variance, 228
Scheduled maintenance, 195
Scheduled service operations, 77
Scope of technical publication, 164
Seasonal product, 94
Semi-finished goods, 9
Sensitivity analysis, 136–37
Series relationship, 50
Shipments, consolidated, 141
SMR codes, 183–84
Source, maintenance and
      recoverability, 183
Spare parts, 10
Spare and repair parts, 23, 175
Spare and repair parts inventory, 92
Sparing decisions, 176
Special rates, 72
Special test and support equipment,
   193
Specified conditions, 49
Stagflation, 7
Stockless production, 121

Storage, 22
Storage costs, 103
Storage facilities, 46
Strategic plans, 225
Student population, 148
Student test material, 158
Subjective probability, 222
Subsistence logistics, 2
Support analysis, 203
Supportability assessment, 204
Surcharge, 71
SV, 228
System, 16
System logistics, 3
Systems concept, 19

T

Table of contents, 171
Tariffs, 71
Task environment, 33
Task oriented, 144
Task oriented objectives, 148
Task times, 153
Technical communicator, 161
Technical publication, 11, 24, 160
   accuracy, 162
   clarity, 162
   organization, 162
   scope, 164
Technical writer, 161
Temporal postponement, 141
Terminal objectives, 149
Test and support equipment, 25, 192
   common, 193
   special, 193
Time phased MRP, 116
Time series lot sizing, 111
Time utility, 83
TL, 71
TM 38-715-1, 177
TM 38-717, 177
To complete performance index, 233

TOFC, 69
Tools of reliability, 50
Top management, 217
Total cost analysis, 4
Traditional management, 218
Trailer on flatcar, 69
Trainee participation, 147
Training, 24
   course length, 145, 152
   performance cycle, 145
   plan, 156
   prerequisites, 147
   requirements, 150
   system, 144
Training materials:
   audio-visual aids, 158
   examinations, 158
   hands-on training guides, 157
   lesson guides, 157
   student text, 158
Trainship, 69
Transfer of knowledge, 144
Transportation, 21, 62
   air, 66
   basic modes, 62
   pipeline, 67
   railroad, 66
   road, 64
   special rates, 72
   water, 66
Trend extension, 41
Truck load, 71
Type 1 uncertainty, 43, 95
Type 2 uncertainty, 43, 95

U

Uncertainty:
   type 1, 43, 95
   type 2, 43, 95
Unscheduled maintenance, 195

User maintenance, 193
Utility:
   place, 83
   time, 83

V

Valence, 222
Validation, 170
Variable cost, 104
Variance:
   cost, 228
   cumulative, 231
   period, 231
   schedule, 228
Verification analysis, 204

W

Warehouse, 34, 81
   economic justification, 86
   facilities, 74
   intermediate-positioned, 86
   market-positioned, 86
   production-positioned, 85
Waterways, 66
Wear out, 52
Wear out failures, 196
Wholesale merchant, 102
Wholesale risk, 96
Work center, 122
Work force, manufacturing, 129
Work year, 80
Workload, average, 119

Z

Zero inventory, 92, 121